The Shorebirds of
North America

The Shorebirds of North America

A Natural History and Photographic Celebration

Pete Dunne

Kevin T. Karlson

Princeton University Press

Princeton and Oxford

Published by Princeton University Press
41 William Street, Princeton, New Jersey 08540
99 Banbury Road, Oxford OX2 6JX

press.princeton.edu

Library of Congress Cataloging-in-Publication Data

Names: Dunne, Pete, 1951– author. | Karlson, Kevin, 1952– author.
Title: The shorebirds of North America / Pete Dunne and Kevin T. Karlson.
Description: Princeton, New Jersey : Princeton University Press, [2024] | Includes bibliographical references.
Identifiers: LCCN 2023045201 (print) | LCCN 2023045202 (ebook) | ISBN 9780691220956 (hardback) | ISBN 9780691224701 (ebook)
Subjects: LCSH: Sea birds—North America. | BISAC: NATURE / Animals / Birds | SCIENCE / Natural History
Classification: LCC QL681 .D887 2024 (print) | LCC QL681 (ebook) | DDC 598.177097—dc23/eng/20231102
LC record available at https://lccn.loc.gov/2023045201
LC ebook record available at https://lccn.loc.gov/2023045202

British Library Cataloging-in-Publication Data is available

PREVIOUS PAGE: *Almost demonic-looking, male Ruffs don an array of eye-catching plumages at lek display areas, with white, black, and rust shadings dominant. During their polygamous lek behavior, these flashy feathers are raised to catch the eye of prospective females. Thanks to Anita North for providing this photo from Norway, taken on May 31.*

Editorial: Robert Kirk and Megan Mendonça
Production Editorial: Natalie Baan
Text Design: D & N Publishing, Wiltshire, UK
Jacket Design: Chris Ferrante
Production: Steven Sears
Publicity: William Pagdatoon and Caitlyn Robson
Copyeditor: Lucinda Treadwell
Jacket images: Kevin T. Karlson

This book has been composed in Brandon Grotesque (headings) and Bodoni Egyptian Pro (text)

Printed in China

10 9 8 7 6 5 4 3 2 1

Contents

Preface

In 1967, a seminal book titled *The Shorebirds of North America* was published by Viking Press, with text by Peter Matthiessen, species accounts by Ralph S. Palmer, and paintings by Robert Verity Clem, all edited by Gardner D. Stout. This work presented a refreshing juxtaposition of writing styles highlighting the imagery of Matthiessen with the natural history approach provided by Palmer. The result is a unique combination of talents that makes this book a must-have for all readers who fancy this interesting group of birds. Unfortunately, it is out of print, which leaves a gap for this type of book on shorebirds.

After a conversation with Robert Kirk at Princeton University Press, the authors decided to attempt to build on and update this original book with one that includes our different styles of writing, but this time with photographs instead of paintings. Pete is the writer who integrates imagery with story lines in the vein of Matthiessen, while Kevin shares Palmer's natural history input. The result is a refreshing celebration of this fascinating bird group.

This book is clearly not an ID guide, nor does it attempt to provide every available bit of biological or natural history information. Instead, it is a textual and visual celebration of the shorebirds of North America that includes a great deal of natural history information, scientific data, and current population numbers and trends for all the shorebirds that breed or regularly occur in North America.

The Shorebirds of North America is divided into two main sections, with the first section sharing general knowledge and information about shorebirds, and the second section, Species Profiles, providing comprehensive natural history information, breeding biology, migratory movements, and population data for every species of shorebird in North America.

A major appeal is the many included photographs. They are not the typical police lineup photos used in field guides, but instead include behavioral images accompanied by descriptive captions, portraits of shorebirds in catchy poses, and many images from Arctic breeding locations never before seen in a published work. Together they provide a memorable visual impact.

Audiences for this book are clearly those who show interest in birds, especially shorebirds, but also the general reader who wants to know more about shorebirds and their natural history. Most information is provided in a casual style, but the Species Profiles contain detailed information about each shorebird species that an interested reader would want to know.

◄ Wilson's Phalarope is a partially aquatic shorebird that breeds in Western prairie locations and winters in high-elevation lakes in the southern Andes mountains of South America. This juvenile was photographed by Scott Elowitz in NYC in August.

Dedication

This book is dedicated to the thousands of shorebird biologists who have indentured their lives to the study of this most fascinating of bird groups, and whose discoveries and insights enliven the pages of this book.

Admittedly this book cannot make amends for all the face-plants into tussock tundra; all the mornings when feet were slipped into boots still wet from yesterday's slog; all the mosquitos fished out of your morning coffee; nor for the clouds of biting midges that descend precisely when rocket nets settle over half a hundred frantic sandpipers. Nevertheless, with thanks and admiration, this book is dedicated to you.

We would also like to offer a special personal tribute to our dear friend and Australian shorebird biologist Clive Minton, who dedicated much of his life to the study, conservation, and protection of shorebirds, and whose untimely death in late 2019 left a gap in our ranks that may never be filled.

On a personal note, the authors would like to thank their wives, Linda Dunne and Dale Rosselet, whose eyes and ears are ever tuned to our works in progress, and whose comments and suggestions are reflected throughout this book. Kevin would also like to add an extra thanks to his wife, Dale Rosselet, for her constant emotional and physical support during his battle with cancer during chemotherapy and surgery in 2021–2022, which occurred right in the thick of the preparation of this manuscript and the photos within.

Clive Minton dedicated much of his life to the protection, research, and love of shorebirds. He worked for several decades as a lead biologist in the banding operations on Delaware Bay, and he leaves a legacy that is both powerful and memorable.

SECTION 1
Shorebirds Overview

Heeding their parents' call to muster, Killdeer chicks are taking cover under soft, protective breast feathers. These highly precocial chicks often leave the nest within an hour or two after hatching and start feeding themselves immediately. CAPE MAY, NJ, MAY

Ask ten shorebird biologists to name the single most amazing trait evidenced by shorebirds and you will probably get ten different answers, all of them defensibly correct. Shorebirds are an acquired taste. Most students of birds pass this group by until later in their avocational development, firstly because not all birders are fortunate enough to live near coastal areas where many shorebirds are concentrated in migration, and secondly because many new birders quail at the thought of separating so many similar species with odd-sounding names that are so unlike the songbirds with which they are most familiar.

However, once birders do embrace this most fascinating of bird groups, they quickly become passionate devotees, plotting their lives around seasonal shorebird concentrations. Shorebirds, you see, are not strictly coastal. Kansas is about as far as you can get from seacoasts, yet the wetlands of Cheyenne Bottoms and Quivera National Wildlife Refuges host two of the planet's greatest concentrations of migratory shorebirds (one in spring; one in late summer/fall).

Residents of land-locked Utah are likewise treated to large seasonal concentrations of shorebirds, including vast numbers of Wilson's and Red-necked Phalaropes that gather on the Great Salt Lake from spring to late summer to feast on the seasonal abundance of brine flies. In 2019, 149,660 Red-necked Phalaropes were counted in aerial surveys from August 16 to 22, and 337,698 Wilson's Phalaropes were tallied from July 11 to 17 (Carle et al. 2022).

Everything about shorebirds is remarkable, from their ability to hide in plain sight to their hemisphere-vaulting migrations. Much acclaim has been accorded a single Red Knot dubbed "Moonbird" because the sum of his 21 years of twice-annual peregrinations from the Arctic to southern South America is more than 400,000 miles, far exceeding the distance from the Earth to our moon (238,900 miles).

But B-95, as the banded knot is identified, is just one among hundreds of thousands of shorebirds of several species whose biannual migrations exceed 18,000 miles each year. Among these is White-rumped Sandpiper, who like Red Knot is a high Arctic breeder and whose migration journeys match those of Red Knot. White-rumped Sandpiper, however, at a mere 1.5 ounces, weighs less than a quarter of Red Knot's breeding weight, and White-rumped has even been recorded in Antarctica, a feat no Red Knot has duplicated.

Another long-distance champion, Bar-tailed Godwit, flies nonstop across the Pacific Ocean in early fall from Alaska to New Zealand over a seven-to-ten-day period, a distance of 7,000–9,000 miles! This amazing feat constitutes the longest nonstop migration of any bird species. Great feats of migration, however, are but one of the compelling facets of a shorebird's life that we intend to explore in this book.

No bird group embraces more breeding strategies than shorebirds. Especially precocial at Stage 2 levels (degrees of independence upon hatching), young Killdeer are able to stand, feed, and even run within an hour or so after hatching.

Foraging mostly in the open, where they are easy prey for hunting eyes, shorebirds use their massed numbers, synchronized maneuverability, and superior flight speed

Red Knot ranks among the planet's migratory champions, as proven by Moonbird, B-95. This bird's rotund body shape reflects a doubling of its body weight in several weeks while feeding on horseshoe crab eggs in Delaware Bay in May. This weight gain is crucial for completing the long remaining flight to its high Arctic breeding grounds.

▲ Tightly packed synchronized flight is how these shorebirds hope to foil the ambitions of hunting falcons as they stage on the Copper River Delta in Alaska in May. These loose murmurations are shorebirds' first line of defense against aerial predators.

▼ While not as brightly colored as orioles and tanagers, shorebirds like this adult male Western Sandpiper do not lack for charm, and in breeding plumage, their tasteful attire is enhanced with touches of rust. This bird is feeding on last year's seeds and plant matter in the high Arctic in late May prior to snow melt.

to outpace predators as swift as Peregrine Falcon and as agile as Merlin.

There are roughly 56 shorebird species that breed or regularly occur in North America, and these are divided into five closely related bird groups: stilts and avocets; plovers; oystercatchers; sandpipers; and jacanas. While many of these are tundra breeders, a goodly number breed or occur at some point of the year across the North American continent. As the name "shorebird" implies, many are bound to coastal or semiaquatic habitats, although several species, most notably Mountain Plover and Killdeer, are dry land specialists.

The boldness of shorebirds invites our affection and their synchronized flight our admiration, but their succulence has at times and in places attracted the attention of gunners and the palates of diners. This gastronomic excess has sadly contributed to the extinction of one North American shorebird, the Eskimo Curlew, a species last documented in 1963 as a bird shot by "sportsmen" in Barbados, where shorebirds are, sadly, still hunted for sport.

As for identification, the widespread Killdeer, a medium-sized plover, ranks among the planet's most easily identified birds. Heck, it even calls out its name. It is a rare birdwatcher who does not account this vocal plover

among the first hundred species on their life list. Found even on athletic fields in urban areas, this bellwether species is the bird that ushers most new birders into the ranks of shorebird devotees.

The authors are fortunate to have lived in shorebird-rich times, while students of birds in the first half of the 20th century were less fortunate, as market gunning decimated shorebird numbers. Pete and Kevin grew up in New Jersey, a state whose mostly unglaciated coastal wetlands teem with migrating shorebirds in spring and fall. What's more, we raised glasses in a time of unencumbered mobility.

We are also fortunate to have studied shorebirds on all coasts as well as the Arctic tundra, North America's great shorebird hatchery, where Kevin gathered many of the stellar images that enliven the pages of this book over the course of four summers as a shorebird biologist. While Kevin's images are found in virtually every treatment of birds published in the last 30 years, nowhere will you find such a rich array as those assembled here, not even in the celebrated book he coauthored, *The Shorebird Guide* (2006).

We are indebted to all those students of birds who preceded us and lent the weight of their insights into this writing, most notably Peter Matthiessen, Robert Clem, and Ralph Palmer, whose seminal book *The Shorebirds of North America* (Stout et al. 1968) inspired this book. So sit back and join us in this visual and informative foray into the amazing bird group that both literally and figuratively straddles evolution's high-water mark. From the fecund reek of Atlantic marshes to the iodine-tainted air that sweeps off the Pacific, and from hot brine-laden puffs of air wafting onto Gulf Coast beaches to the ice-cap–chilled "no-smell" Arctic air, there are shorebirds to enliven the landscape and kindle the fascination of poets and biologists alike.

A Leap of Faith: A Shorebird Spectacle

Somewhere in the skies off the Atlantic Seaboard, flying at 60 miles per hour and at an altitude of 15,000 feet to reduce drag and keep from overheating, the tired birds use up the last of the fat reserves they had laid down on the beaches of northern Brazil. But the Red Knots have no choice but to continue on to the beaches of Delaware Bay, where on the high tides of May, they hope to find a bounty of lipid-rich horseshoe crab eggs to replenish their exhausted fuel supply and carry them on the final leg of a long journey to their breeding grounds.

Bivalves are the normal grist for the Red Knot's industrial-strength digestive tract, but with their guts atrophied to reduce weight for their long flights, these migrating birds seek out and digest the nutritious soft-bodied eggs of horseshoe crabs, bird pabulum. Until their internal organs are restored, the birds are temporarily dependent on the bounty of eggs. It is Delaware Bay or nothing, and now it is a race against time.

Some migrating shorebirds double their body weight with fat reserves prior to migration, but they must strike a

balance between cargo and aerodynamic design. Every added gram means more weight to carry aloft, and too much ballast can compromise their drag-reducing profile. It is a delicate balance between the need for fuel and for efficiency, but with good flying conditions and a modest tail wind, the fuel carried by knots is sufficient to bring them to their refueling depot in Delaware Bay. By the time the birds start their descent, many are metabolizing muscle tissue, burning the very engines that bear them aloft in a last-ditch bid to reach the egg-rich shores where the feeding frenzy is already in progress.

Their wingtips nearly touch the sand as they glide down above the glistening ranks of mating crabs and the frenzy of feeding Sanderlings, Ruddy Turnstones, and Semipalmated Sandpipers, and the famished knots begin knocking back the tiny eggs deposited in gray-green lines by the lapping waves.

It is a frenzy of sex and gluttony—overturned crabs waving their legs skyward while sandpipers dart among the stranded crabs and loft into the sky in a spray of birds every time a large wave washes ashore. Laughing Gulls raise their raucous voices over the crockery-like clatter of surf-churned crabs, and adding to the din is the clamorous chatter of sandpipers. This frenetic ribbon of life stretches along both sides of Delaware Bay and involves more than a million and a half migrating shorebirds and tens of thousands of crabs, all driven into a frenzy to feed and breed.

If conditions are right, the knots will nearly double their weight to 220+ grams in about two weeks and set forth on the final leg of their journey to the high Arctic breeding grounds of the Canadian Archipelago. If they arrive in prime breeding condition (i.e., retaining adequate fat reserves), the birds will court, breed, hatch young, and then begin their return journey to southern South America by late July (sooner for failed breeders and females). It is a quick turnaround, but summers in the high Arctic are brief, with snow returning for the winter by early to mid-August. Successful breeding for this subspecies of Red Knot is utterly dependent on the food resources of Delaware Bay.

Amazingly, this great shorebird concentration was unknown to science when that first version of *The Shorebirds of North America* was published in 1967. Now it is accounted as one of the greatest wildlife spectacles on the planet, and one that at least four species of shorebirds

A flock of Red Knots descends to join the feeding frenzy on horseshoe crab eggs on the shores of Delaware Bay each May. Their fat reserves depleted, these migrating Red Knots are dependent on the bounty of horseshoe crab eggs laid down by the waves to double their weight in several short weeks.

A cross section of the frenetic ribbon of northbound shorebirds that crowd the beaches of Delaware Bay in May shows Red Knots, Ruddy Turnstones, Sanderlings, and Semipalmated Sandpipers, along with ubiquitous Laughing Gulls. These four shorebirds and Laughing Gull compose most of the feeding birds at this important migratory stopover location.

rely on regionally for all or part of their breeding success and survival. But it is not the only great gathering of shorebirds on this continent, and a timely arrival and departure is only one of the high-stakes games that shorebirds play with the universe.

The Perfect Bird

Challenged to design the perfect bird, we would without hesitancy delve into ranks of shorebirds and reach down to the Scolopacidae (sandpipers), the largest of the roughly six or seven family groups in the shorebird order Charadriiformes. Choosing any one of North America's 37 regular breeding sandpipers, it is doubtful we could contrive any refinements that would enhance their array of superior attributes.

With their long legs and splayed toes, shorebirds are ideally suited to exploit the food riches that flourish where land and water meet. Their extraordinary powers of flight

permit some of them to vault entire hemispheres in a single bound, placing seasonal food abundance at opposing ends of the earth within reach of their tweezer-like bills. Designed with football-shaped bodies and scimitar-like wings, few bird families can match the speed of shorebirds, and no other bird group can equal their nonstop migrations.

As for staying power, shorebirds have long stood the test of time. Proto-shorebirds with sandpiper or plover-like feet were foraging stride for stride beside their non-avian dinosaur relatives along ancient lakeshores during the early Cretaceous period (145–65 million years ago). The proof is etched in tracks preserved in 122-million-year-old rock. Those nonfeathered members of the dinosaur family tree reside in museum specimen trays today, but the planet's ever-changing 217 or so shorebird species endure (WHSRN 2021).

Admittedly, shorebirds are not as colorful as some might envision their "perfect bird," but being cryptic suits a bird group that forages in open habitats beneath the rapacious gaze of multiple falcon species. As for bird song, one of the

◀ This Hudsonian Godwit shows all the physical attributes that characterize shorebirds as "The Perfect Bird": football-shaped body, scimitar-like wings, long legs, and splayed toes. He is surveying his breeding territory from the top of a stunted spruce tree at the intersection of boreal forest and tundra in Churchill, Manitoba in June. A gust of wind caused him to raise his wings, allowing Kevin to get this special photo of one of his favorite birds.

▶ The haunting cry of the curlew ranks among the most inspiring sounds in nature and is in fact the source of the bird's name. Long-billed Curlew is our largest shorebird, and its long bill allows the bird to deeply penetrate moist substrates to find invertebrates that lesser shorebirds cannot reach. TEXAS, SEPTEMBER

most celebrated attributes of birds, shorebirds certainly take a back seat to the warblers and thrushes. But shorebirds are far from mute, and the wild cry of the curlew and the plaintive whistle of Black-bellied Plover rank among the planet's most evocative sounds.

For human beach strollers, the conversational banter of feeding sandpipers is as much a part of the shore experience as the tug of surf on ankles and the tang of salt-laden air. For as long as humans have walked ocean shores, shorebirds have delighted our senses with their frenetic movements, wheeling flocks, and lively banter.

While gulls, terns, and auks are also included in the order Charadriiformes; these bird families are not typically referenced as "shorebirds," a designation reserved for the oystercatchers, stilts, avocets, plovers, and sandpipers. Throughout the rest of the world, however, shorebirds are referred to as "waders." By whatever name you know them, we are delighted to offer this tribute to the bird group admired by students of birds worldwide.

Superlatives notwithstanding, some generalities apply to most shorebird species. For example, most shorebirds are highly mobile, covering extensive areas when foraging. Many are also gregarious outside the breeding season, massing into flocks that number from a few birds in some species to tens of thousands in others. On their Arctic breeding grounds, shorebirds are widely spaced over tens of thousands of square miles. Shorebirds are also carnivorous, feeding primarily on small animals, and many species are associated with mostly wet environments.

Anyone who doubts the evolutionary link between birds and dinosaurs has never watched a Black-bellied Plover forage across a tidal flat. As sight and sensory hunters propelled by strong legs that permit rapid acceleration, these dedicated carnivores with bullet-shaped bills feed voraciously on marine worms secured after a calculated stop and go approach, a short dash, and a one-sided tug of war.

With their large heads and quasi-upright stances, these King Plovers even somewhat resemble Tyrannosaurus Rex, arguably the most formidable predator ever to stalk the earth. While mostly solitary, as befits a top-tier predator, Black-bellied Plovers may gather during migration in loose flocks of up to a hundred birds to forage in freshly plowed, worm-rich agricultural fields. While primarily visual feeders, plovers also locate underground worms by feeling their slight motion with the sensitive soles of their feet.

Other shorebirds are tactile feeders, probing beneath easily penetrated substrates (typically wet) in search of invertebrate organisms. Always on the move and often clustered in tight-packed flocks, birds surge like a hungry wave across open, food-rich environments, ever probing for buried prey. While called "sandpipers," most are actually mudpipers, favoring this easily penetrated substrate over sandy flats whose surface, once shed of a film of water, compacts into bill-defeating concrete. Species that forage on sandy beaches typically do so at the water's edge, where sand is still in semi-suspension and easily penetrated by a thrusting bill.

Much can be deduced about a shorebird's feeding habits and its prey by studying bill length and shape. Small sandpipers have short, pointed bills well suited for picking or abbreviated probing, while the larger curlews and godwits have long, scimitar- or lance-like bills designed for deep penetration. Avocet bills are scythe-like and designed for sweeping the water column, while the straight, sturdy bills of oystercatchers are configured for

◀ No one who has ever watched Black-belled Plovers foraging across tidal flats can gainsay the genetic linkage of modern birds and dinosaurs. This attractive breeder is one of the most globally widespread of all shorebirds and is found on all continents except Antarctica. NJ, MAY

▼ While most sandpipers are really "mud" pipers, the aptly named Sanderling prefers sandy substrates for picking and light probing. This photo reflects the stark beauty of black, white, silver, and gray tones with a high-contrast background as a juvenile Sanderling stitches the sand while foraging. NJ, SEPTEMBER

chiseling carapace-encased organisms off rocks or for decanting mollusks. Plovers have short, sturdy bills suited for plucking prey from dry surfaces or light probing, and phalaropes have needle-like bills tailored for picking or straining water droplets for microscopic prey.

It is not merely massed numbers of birds that humans find so inspirational about this family group, but the great leaps of faith exhibited by these long-distance voyagers as they set out twice each year into the unknown, guided by instinct but subject to the perturbations of an indifferent to hostile universe.

Shorebirds, of course, did not simply spring full blown from the surf readily equipped to vault hemispheres, pack on weight, and arrive precisely in time to exploit food riches at opposing ends of the planet. They are the product of millions of years of evolutionary push and pull. It's Darwin meets Kierkegaard, clad in feathers and divided into roughly 52 North American breeding species (WHSRN 2021), each with a story to tell about the tiny birds that could, can, and continue to meet the challenges of an ever-changing planet.

No, there is no one single overriding trait that makes shorebirds the superior creatures they are, but rather the compounded array of superior traits that propel shorebirds to the apex of evolutionary development and the height of our esteem.

Habitats

It is no more accurate to say that most North American shorebird breeders are tundra nesters than it is to say that most shorebirds are coastal migrants. True, more than half of North American shorebirds are Arctic breeders, but the tundra environment is as varied as the coastal environment. Some tundra breeders like Sanderling and Red Knot select the stony, sparsely vegetated habitat of the high Arctic, while others like Red Phalarope, Semi-palmated Sandpiper, and Ruddy Turnstone are drawn to the wetter, coastal grassy Arctic tundra that characterizes much of the North Slope of Alaska and Canada.

Still others like Whimbrel and Bristle-thighed Curlew favor drier, grassy hilltops in more upland tundra habitats. Semipalmated Plover breeds on gravel bars along watercourses as well as adjacent open areas, and the large plovers favor drier, raised areas. Hudsonian Godwit and Short-billed Dowitcher breed in the wet ecotone where tundra and taiga forests meet, perching atop stunted trees while surveying their territory.

And some shorebirds breed in bogs within the vast northern forests, including Wilson's Snipe, Greater and Lesser Yellowlegs, and Least Sandpiper. Others are not northern breeders at all and include prairie specialists like Long-billed Curlew, Upland Sandpiper, Western Willet, Wilson's Phalarope, Snowy Plover (Great Basin race), and Marbled Godwit.

A handful of shorebirds actually do breed on or adjacent to coastal shores; these include American and Black Oystercatchers, Piping, Wilson's, and Snowy Plovers, and Eastern Willet. Only one species, Northern Jacana, is wholly wedded to an aquatic environment. Placing their nests on floating vegetation and relocating as water levels demand, these tropical and subtropical shorebirds are even able to walk on water (more or less) (see page 269). Some shorebirds like Killdeer and Mountain Plover are dry land specialists that eschew wetlands. Another widespread eastern sandpiper, American Woodcock, nests in the leaf litter of deciduous woodlands and winters in damp, earthworm-rich forested environments, often south of the freeze line.

In migration and during the nonbreeding season, many shorebirds do live up to the implications of their name. But here too, not every coastal environment is the same, nor do all shorebirds favor the same coastal habitats. Many of the smaller sandpipers, or "peeps" as they are colloquially known, fan out across mudflats exposed at low tide, and only a handful of nimble or long-billed species are able to exploit the food resources of sandy coastal beaches with dynamic wave action. These include Sanderling, Semipalmated and Western Sandpipers, Red Knot, Marbled Godwit, American Oystercatcher, Eastern and Western Willets, and Long-billed Curlew.

Many shorebirds are probing species whose long legs and long bills allow them to access and penetrate the substrate of shallow wetlands. These include Long and Short-billed Dowitchers, Dunlin, Whimbrel, and Stilt Sandpiper. While Long-billed Dowitcher favors freshwater wetlands, Short-billed Dowitcher prefers tidal or brackish coastal areas away from breeding grounds.

One species, Solitary Sandpiper, shuns coastal areas as much as it avoids crowds. This methodical feeder, closely related to the yellowlegs, searches shallow, freshwater puddles or ephemeral wetlands for invertebrate prey often overlooked by less diligent hunters.

In tidal areas, shorebirds need higher dry land or shallow standing water where they can wait for high tide to recede. These roosts require a measure of solitude away from predators or human activity. Only two species, Red-necked and Red Phalaropes, forage in offshore marine environments, and they are the only shorebirds that can float in open water for extended periods (see page 266, top photo). Other shorebirds reluctantly forced to land in the water at sea, other than Wilson's Phalarope which can swim in water for hours on end, must quickly return to the air or become waterlogged. However, American Avocets frequently forage belly deep in the saltwater of the Gulf of Mexico for extended periods, and their webbed feet allow them to swim in deeper water if necessary.

In migration, many shorebirds are able to exploit man-made habitats. In the Midwest, fallow agricultural land has supplanted the buffalo-grazed prairie once used by migrating American Golden-Plover, and Mountain Plover has shown a tendency to nest in tilled land in addition to the adjacent high prairie it calls home. Plovers also utilize fallow agricultural fields for foraging in winter.

Moist sod farms and golf courses regularly attract migrating Buff-breasted Sandpipers, which find the close-cropped vegetation approximates the short-grass habitats they favor. Other friendly habitats for Buff-breasted Sandpiper include cemeteries, rice fields, airports, and recently harvested fields. American Golden-Plover and Pectoral

◁ In the taiga habitat where they breed, Short-billed Dowitchers perch on the tops of small, stunted trees to survey their breeding territory and look for potential danger from an elevated position. This behavior of sandpipers perching in trees is rarely seen away from breeding grounds, both Arctic and temperate. CHURCHILL, MANITOBA, JUNE

△ Red Knot and American Oystercatcher are two shorebirds that forage in the active wave zone that other smaller shorebirds avoid. In this case, both are searching for mole crabs, a high-protein, soft-shell crustacean that lives just beneath the sand's surface, and whose presence is revealed by air bubbles as waves retreat. These six juvenile Red Knots with one adult on August 30 in NJ suggest a productive high Arctic nest season.

▷ Buff-breasted Sandpiper is a delicate and attractive shorebird that frequents open grasslands and wet meadows during migration. It has a certain indefinable mystique that makes it one of the most sought-after shorebirds, and it has always been Kevin's favorite bird in the world. NY, SEPTEMBER

Sandpiper also find these habitats attractive and productive for feeding during migration.

In drier areas, water and sewage treatment facilities are meccas for migrating shorebirds, and some progressive municipalities control water levels for them. In the Mississippi Delta, a busy migratory corridor for inland shorebirds, some local rice farmers cooperate with the Delta Wind Birds group of the Mississippi Ornithological Society by lowering water levels in some rice fields and flooded catfish pools during late summer and fall migration to provide critical feeding areas for shorebirds of many species.

Life Spans

Life spans of shorebirds vary greatly, with some smaller species living on average 3–10 years (Least and Semipalmated Sandpipers 8–10 years), while some larger species, like Black-necked Stilt and Whimbrel, may live 20 years or more. If juvenile birds survive their first year, life spans greater than ten years are not uncommon for some shorebird species (Plauny 2000).

Some species in the genus *Numenius* (curlews) can live from 20 to 30 years, with the oldest banded pair of Eurasian Curlews living 32 years. The oldest recorded Whimbrel was 24 years and 1 month old based on banding data, but a life span of 10+ years is more typical of this bird group (birdfact.com, n.d.).

Predator Defense

When shorebirds are sighted by a predator's eyes, the birds employ two of their most celebrated predator defense mechanisms: rapid flight and close air maneuvering. Many eyes see danger better than two, and flocking birds such

as shorebirds exploit this advantage by taking to the air as danger approaches, not when it is already upon them. Open sky is their refuge, since shorebirds enjoy aerial parity or supremacy over feathered predators that rely mostly on surprise to their advantage in the age-old contest of predator versus prey. Once airborne, a shorebird flock coalesces and climbs, wheeling in a synchronized, tightly packed mass that confuses hunting eyes. Through sheer numbers and shifting possibilities, the ability of a predator to concentrate on a single target is nullified.

The escaping flock also signals its distress to other birds in the vicinity by alternately flashing counter-shaded light underparts and darker upperparts, which appear like puffs of smoke on the horizon, a smoke signal to other members of the tribe that reads, "heads up, hungry falcon here."

Spiraling skyward, the flock either outlasts the hunter or perhaps sacrifices an injured or weaker member, thus preserving the fitter members of the flock and benefiting their genetic pool, moving the perfect birds closer to absolute perfection.

Shorebird Plumages

Of all the biological refinements that permit shorebirds to exploit the opportunities of an open environment, none is more understated than plumage, the cryptic enfolding of browns, buff, and gray with touches of black and white that permit shorebirds to disappear in plain sight. This same subtle color combination predominates in private clubs,

◄ It may seem like the picture of serenity, but this incubating Whimbrel near Churchill, Manitoba in June is on high alert, the first and last line of defense between its clutch and marauding predators. Whimbrel is so large that it cannot hide in the close-cropped tundra, so it keeps a constant vigil for danger and potential disturbance.

▲ Tightly packed, synchronous wheeling flocks (murmurations) are a shorebird defense strategy used to distract aerial predators like Peregrine Falcon and Merlin. The hypnotic, swirling motion of these birds (mostly Dunlin and Sanderling) confuses birds of prey, which need to separate single birds from the pack for successful hunts. NJ, MARCH

◄ From above, this dapper Baird's Sandpiper would be nearly invisible to hunting eyes. Were a falcon to approach, the bird would crouch, eliminating the tell-tale shadow. Cryptic plumages like this allow shorebirds to survive in open spaces just by blending in with surrounding environs. TEXAS, APRIL

where gentlemen are required to wear jackets to dinner and tweed predominates.

The cryptic nature of shorebird plumages was brought home to Pete in the following instances:

STORY 1—Conducting aerial surveys along Delaware Bay, 1982

Flying 50 yards offshore, we leveled out at 100 feet, with Moore's Beach, a shorebird hotspot, filling the plane's windshield. At 200 yards out, the beach had disclosed none of the dark and light shading that would have indicated clusters of birds. The beach showed only bland uniformity, so "totally empty," our survey team surmised. But the beach couldn't be empty, as it was May 27, the very height of migration; however, if not empty, then the entire 2-mile stretch of sand would have to be carpeted with birds from marsh to water's edge!

Nothing in our evolving survey protocol had prepared us for such a challenge! As the distance narrowed and the beach erupted in wings as tens of thousands of sandpipers

sought the safety of open air, we were forced to accept a reality that stood apart from any we knew. THERE IS NO VISIBLE BEACH; THERE ARE ONLY BIRDS HERE.

As the plane sped by, it seemed for all the world as if the earth was reaching for the sky. It was the job of survey coordinator Wade Wander to count, and Pete was charged with establishing a species breakdown. They did the best they could, but their resolve passed through the gyrating mass like bullets through smoke.

"Back one time again, Hank?" they asked the pilot. Swinging out over the bay, letting the birds settle, they tried again. Staying farther offshore this time and knowing what they were up against, the team did better. Cutting the rising ribbon of birds into 1,000-bird sections and relying on Pete's eyes instead of binoculars to assess species composition, they came up with figures that inspired confidence.

"Seventy thousand birds on that stretch of beach," Wade announced. "Make it 75% Sanderling; 20% Ruddy Turnstone; balance peep," Pete intoned. The birds settled in our wake and melted into the sand.

▽ As feeding shorebirds lift off the beaches of Delaware Bay, it seems as though the earth is reaching for the sky. In a short time, this spectacular concentration will be apportioned across their Arctic breeding grounds. This photo captures the frenetic, fast-paced activity at this location in May, with Red Knots, Ruddy Turnstones, and Sanderlings the main players.

STORY 2—American Golden-Plovers

Pete approached a nesting American Golden-Plover whose scrape was on a raised mound in the Canadian Archipelago. Stopping where he judged the bird to be, his eyes could discern nothing on the stony, lichen-encrusted hilltop until the bird irrupted at his feet, disclosing her precious clutch of three grayish, brown-spattered eggs, as indistinguishable from the tundra as the adult in breeding plumage.

STORY 3—Nearly Invisible Plovers

During the 2007 World Series of Birding competition, Pete's team was running behind schedule and arrived at the South Cape May Meadows at dusk. Still needing Piping Plover for their list, they trained spotting scopes on the center of the 2 × 2 ft. wire exclosure protecting the incubating bird from predators. "But in the fading light, the sand-colored bird crouched in sand and defeated our eyes. We were forced to admit that none of us could see a bird, even though it was certain we were looking right at it."

Piping and Snowy plovers' pale plumages are almost invisible against dry sandy beaches, while the darker upper parts of Semipalmated and Wilson's plovers are the color of dirt or wave-washed sand, making them nearly invisible in the wave zone where they forage. Mountain Plover's upperparts match the color of prairie grass and bare earth, and the double breast-bands of Killdeer help break up the starkness of the bird's white underparts.

Shorebird eggs, like these of American Golden-Plover, are unusually large to allow precocial chicks to develop inside the egg. The cryptic camouflage of these eggs against the Arctic tundra renders them virtually invisible to nearby ground or aerial predators. ALASKA, JUNE

Many shorebirds have breeding and nonbreeding plumages, and while rarely flamboyant, those sporting touches of rufous and gold add to their visual appeal. Some shorebirds, however, have bold breeding plumages, with the harlequin pattern of Ruddy Turnstone simply striking, and the stunning bright rust underparts and contrasting black-and-white head and gold-edged back of Red Phalarope unforgettable. Mostly it is the males of species that don brighter plumages for breeding, but among phalaropes, a genus that allocates the entire nest duties to males, it is the

Red Phalaropes don their colorful breeding plumage for a handful of crucial months. This attractive female is as buff as they come and almost certain to catch a suitor's eye. Female phalaropes are the dominant member of the pair bond, and the most strikingly plumaged as well.
ALASKA, JUNE

For much of the year, the aptly named Dunlin is clad in cryptic, dun-colored plumage. But in spring, males and females get into their breeding attire, which is more eye-catching than its drab nonbreeding plumage. Note the variability of Dunlin's winter plumage (*right*), especially the complete brown background on the left bird's head and breast. CHURCHILL, MANITOBA, JUNE (*LEFT*); NJ, FEBRUARY (*RIGHT*)

female that is more strikingly garbed. Outside the breeding season, phalaropes and most other shorebirds are mostly gray or brown, with some variation for habitat preference.

A prime example of seasonal plumage variation can be seen in Dunlin, whose breeding plumage for both male and female shows a rufous back and black belly, but who present a bland dun-colored plumage the balance of the year. Where other bird families shout with their attire, shorebirds whisper. Consequently, shorebirds are able to exploit food riches where tanagers and orioles cannot, the fecund shores that stretch across the planet, exploitable by a bird group that can disappear in plain sight.

Foraging and Feeding, or "Why shorebirds seem ever on the run"

Shorebirds as a whole and sandpipers in particular exhibit an array of specialized feeding techniques, although their invertebrate prey is fundamentally similar. During the breeding season, adults and chicks feed primarily on insects and spiders, which juveniles procure themselves. Outside the breeding season, many species switch to a diet of marine and freshwater invertebrates, most notably bivalves, Polychaeta, small crabs, marine worms, and for some species, small fish.

A good number of shorebirds forage in 4 inches of water or less. Some are specialized to feed on dry land, and only a handful are equipped to exploit the food riches of deeper water, with two phalarope species (Red and Red-necked) able to utilize the riches of the ocean environment. To avoid competition, shorebirds engage in "resource partitioning," with different species using dissimilar foraging techniques and water depths to secure insects and larvae, marine invertebrates, small bivalves, and small fish.

Almost all shorebird feeding in coastal environments is subject to the tides. Birds forage when it rises and falls, and rest when it peaks. Unlike plovers, most of these tactile foragers are flock feeders, sometimes moving shoulder to shoulder across invertebrate-rich environments. Falling water levels give probing species, including dowitchers and Dunlin, easy access to the invertebrate riches below the surface of the mudflats.

Above the waterline are the plovers, which are mostly sight hunters. Plovers are largely land predators, feeding away from other plovers and standing tall to sight prey wholly or partially exposed on the surface. Walking in a coursing pattern, foraging plovers pause every few steps before racing to capture sighted prey, which they grasp with strong bills, or in the case of worms, exhume after a one-sided tug of war (see page 87, lower photo). Many plovers also employ foot pattering in shallow water or wet sand, a technique that induces prey to move so the rod-cell–enriched eyes of plovers can spot them.

A plover's methodic walking motion when foraging often causes prey to move just under the surface, which

◀ A few plover species use this foot-pattering behavior to cause invertebrate prey in wet or moist substrates to move around or come near the surface. This juvenile Semipalmated Plover feels or sees its prey and probes the sand to grab a meal. NY, AUGUST

▼ Members of the "rockpiper guild," a nonbreeding Surfbird from California in November (*left*) and breeding Wandering Tattler from Alaska in May (*right*), are nimble and sure-footed on slippery rock substrates. These medium-sized shorebirds spend most of their time away from breeding areas on Pacific rocky shorelines or jetties.

plovers sense with the sensitive soles of their feet. They stare at the substrate where the movement was felt and then probe the general area until the prey is secured. This is similar to the way American Robin finds worms under the surface of the ground. Now comes the hard part, defending captured prey from the pilfering predilections of other plovers or gulls. Worms are ingested whole, while small crabs require a bit of mastication to reach the succulent inner body.

But it is the sandpipers where we find the greatest array of foraging techniques, with these directly linked to their bill length and shape. Above the waterline are a handful of sure-footed sandpipers ("rockpipers") which are specialized to exploit the intertidal riches within the splash zone of rocky coastlines. These "rock specialists" include Surfbird, Wandering Tattler, Dunlin, Ruddy and Black Turnstones, and Rock and Purple Sandpipers. Scrambling over slippery rock surfaces, these sure-footed acrobats explore exposed tiny mussel beds, kelp, and algae with probing bills for marine invertebrates, or they may swallow small bivalves and sea snails whole to be pulverized in their gizzards.

On the Pacific Coast, these splash zone feeders may be joined by Black Oystercatcher, which uses its chisel-like bill to secure limpets, periwinkles, and chitons as well as bivalves, whose partially open shells invite the bird's thrusting bill. American Oystercatcher will also visit rocky jetties in winter to feed on small mussels.

Below the high tide line and within the wave zone are larger long-legged and long-billed probers that explore the wave-churned substrate with probing bills in search of crustaceans and marine invertebrates. These wave zone specialists include curlews, willets, and godwits. American Oystercatchers and Red Knots are two more frequent hunters in this ecozone, where they probe the sand after the waves retreat in search of one of their favorite high-protein foods, the soft-shelled crustacean mole crab. Also found in this dynamic zone are Sanderlings, a sandpiper that plays tag with the waves and probes the wave-stirred sand within the receding wash with stitching bills.

Out in the calmer standing water and shallow flooded flats are mud probers like Dunlin and dowitchers that move in clustered ranks, driving long bills downward into the substrate with metronome regularity. Longer-legged

Unlike most shorebirds, young American Oystercatchers enjoy a prolonged apprenticeship under adult care when learning how to forage for crabs and to crack open crustaceans with powerful, laterally compressed bills. This juvenile 3-month-old "teenager" whose bill is still growing just accepted a mole crab from one of its parents. NJ, LATE JULY

Stilt Sandpipers also plumb these depths, but they prefer freshwater habitats and probe the water with a series of rapid jabs, often submerging their heads when doing so.

Farther out are Greater and Lesser Yellowlegs, active sight hunters that respectively stab and pick at prey within the water column. Farthest are Marbled Godwit, Black-necked Stilt, and American Avocet, whose long legs and long bills permit them to capture prey at depths beyond the reach of other shorebirds.

Phalaropes pick both insects and their larva from the water column and strain microscopic organisms from water droplets with their needle-like bills. Wilson's Phalarope commonly races along lakeshores after flies and other insect prey, even snapping prey out of the air. All phalaropes engage in spinning on the water's surface, a maneuver designed to ferry prey to the surface via the resulting water vortex.

When the tide retreats, buried invertebrates are within reach of the bills of the smaller sandpipers. The tiny birds carpet the mudflats, probing the shallows and exposed mud or skimming the riches of the mucus-like broth (biofilm) that coats the exposed surface of tidal flats. Tiny Least Sandpipers prefer drier areas just above the water line, while burly White-rumped Sandpipers, feeding up to the gunnels, forage in water often too deep for Western and Semipalmated Sandpipers to exploit. Some birds defend small territories within these large feeding areas, but challenges are few if prey is abundant. The advantage of having many eyes watching for danger also mitigates any disadvantages associated with competition.

Like a needle-billed whirling dervish, this breeding female Red-necked Phalarope spins around to create a vortex in the water column that ferries insect larvae and invertebrates to the surface, where they are snatched by the often-aquatic shorebirds. ALASKA, JUNE

Western Sandpipers feed heavily on the mucus-like, nutrient-packed biofilm that coats mudflats when the tide recedes. This biofilm is made up of numerous microorganisms that stick to each other and to the mud's surface. TEXAS, APRIL

Some days you're the woodcock, some days you're the worm! American Woodcock locates its favorite earthworm food by inserting its long bill into the soil while moving the flexible tip around. When the worm retracts from the bill tip's motion, the woodcock feels this with the sensitive tip and grabs its meal under the surface. NJ, FEBRUARY

And some shorebirds are simply not tied to aquatic environments. These terrestrial specialists include Buff-breasted Sandpiper, Mountain Plover, Upland Sandpiper, and Killdeer, species adapted to capture insect prey in dry open or grassy terrain. Fondly called "grasspipers" by seasoned birders, these prized birds are often joined by other shorebirds in moist or flooded areas after rains, including Pectoral Sandpiper, Black-bellied Plover, American and Pacific Golden-Plovers, Long-billed Curlew, and Baird's Sandpiper.

Not particularly impressive until seen in action are the wedge-shaped bills of turnstones, which, in coordination with their strong neck muscles, lift and toss shells and other debris off beaches, exposing hidden prey beyond the leveraging capacity or ingenuity of other beach feeders. Ruddy Turnstone are fundamentally omnivorous and often excavate body-concealing craters in sand to reach the buried egg masses of horseshoe crabs. They are also found begging for fish scraps at fishing docks or snatching dropped food at beach picnics in southern and tropical areas in winter.

American Woodcock avoid competition by using their long, Popsicle-stick bills to probe moist upland forest floors and fields, habitats shunned by other shorebirds. Using their touch-sensitive tweezer-tipped bills, they reap a harvest of buried earthworms that would stir envy in American Robin, a dietary accomplishment that led to woodcock's partially digested prey to be prized by Victorian epicures who regarded "trail," as it was called, a fitting side dish to gamebird courses.

While most shorebirds are able to forage in salt or freshwater–washed environments, one species, Solitary Sandpiper, is mostly a freshwater obligate. This methodical feeder may spend hours canvassing a puddle the size of a bathtub, securing prey that is overlooked by less methodical feeders.

Of course, the shorebirds that most commend themselves to scrutiny are those whose bills gave rise to specializations beyond the grasp of the competition. Notable features among these longer-billed species are the forceps-like bill of Long-billed Curlew and the lancelike bills of godwits and willets, all designed to probe deep and extract hidden prey beyond the reach of lesser birds.

Whimbrel and Long-billed Curlew have curved bills that are ideally calibrated to extract fiddler crabs from their subterranean chambers that are curved to foil birds with straight bills, and the chisel-like bills of oystercatchers are hardened to pry chitons off their rocky base or decant mollusks with blunt trauma or surgical precision, similar to an oyster-shucking knife.

△ Shorebird bills are specialized for different prey and hunting techniques. Can anyone doubt the bill of Long-billed Curlew was made for probing? This species can access food well below the surface, beyond the reach of most other shorebirds. TEXAS, SEPTEMBER

Whatever method they use or whichever prey they seek, shorebirds must consume about one-third of their body weight per day to meet the energetic needs of their high metabolism. They require even more food if their objective is to lay down body fat to fuel migration. It is precisely this metabolic race with resources that explains why shorebirds seem ever on the move, feeding constantly and relocating to places that offer greater promise. It is also why resource partitioning is so important to reduce competition.

At times of opportunity, some shorebirds become mostly vegetarian, gorging themselves on ground-ripened berries or seeds to enhance laid-down fat reserves, or to see them through insect-impoverished times. These intermittent vegetarians include Whimbrel, American Golden-Plover, and the extinct Eskimo Curlew. Northern Jacana also ingests vegetable matter, but whether this is intentional or a byproduct of their lily pad–plucking foraging style is unknown.

So, why do shorebirds hurry or run while foraging? To keep ahead of metabolic demands that are greater than those of many other bird groups.

Courtship and Breeding

Overview

Shorebirds are one of the most efficient and successful bird families on earth, and their survival over countless eons is a testimony to this success. For thousands of years, their numbers remained steady, and many species even increased because of their ability to deal with natural dangers, including predators and a changing climate. A heightened sense of awareness evolved with their open-space lifestyle, allowing them to quickly perceive potential dangers in the form of ground or aerial predators, and several strategies and behaviors both during and after the breeding season have resulted in a better than average survival rate. For birds that spend most of their lives without vegetative or geomorphic protection, it is important that they successfully produce ample numbers of offspring that can quickly adapt to their nomadic wandering lifestyle.

Shorebirds are not mammals. They belong to a group of warm-blooded vertebrates constituting the Class Aves, characterized by feathers, toothless beaked jaws, the laying of hard-shelled eggs, a high metabolic rate, a four-chambered heart, and a strong yet lightweight skeleton. All birds, including shorebirds, are actually feathered Theropod dinosaurs and constitute living dinosaurs in today's world.

Courtship

Shorebirds employ a variety of mating strategies to move their genetic dowry onward. Some bird species signal their genetic superiority with vivid plumages, others with song. Shorebirds mostly take to the air. After establishing territories, males engage in energetic flights that map the sky to demonstrate to prospective mates their health and breeding fitness.

A few shorebirds like Buff-breasted and Pectoral Sandpipers are lekking species, where multiple males vie to catch a female's eye with ritualized displays. Other species simply go ballistic, weaving patterns in the sky accompanied by strange but encouraging sounds. A perfect example is American Woodcock, a widespread forest shorebird, whose reedy "peent" on still spring evenings occurs across

Buff-breasted Sandpiper has one of the most complex courtship and mating rituals of all birds. This double-wing embrace posture is assumed by the polygamous male after prospective females are attracted to his single-wing display. It is the last step before one or multiple females approach him to compete for copulation. Males provide no further help with incubation or brood rearing. ALASKA, JUNE

Solitary Sandpiper is one of only two shorebirds in the world (Green Sandpiper from Eurasia is the other) that use abandoned songbird nests to lay their eggs. The nests of American Robin, Canada Jay, Rusty Blackbird, Eastern Kingbird, and Cedar Waxwing are the most used. This rare photo of a bird sitting in an American Robin's nest is a scene almost never encountered in the vast regions of their boreal forest breeding range because of their secretive nature and the height of some nests. Ted Swem found this bird nesting close to his home near Fairbanks, Alaska in early June.

much of eastern North America, the sound of the winter-stilled landscape's reawakened pulse.

Peent…peent…peent, then up, up, up, and up goes the bird, only to spiral down in a cascade of twittering notes produced by their outer primary flight feathers and a touchdown in some open area close to the forest edge, where multiple males often display in close proximity. A female selects her champion, and then begins the nesting process alone. Males continue to display, often breeding with multiple females over the course of the breeding season that may last four months.

Other shorebirds are mostly monogamous, with males or females or both incubating the three or four eggs, and one or both parents tending to the hatched young. Females of several species, including Spotted Sandpiper, may leave their first clutch to the care of the male and subsequently mate with a second male to produce another clutch of eggs. Otherwise, a single clutch of eggs per nesting cycle is typical of shorebirds, which given the relatively short life span of some smaller species (average 3–10 years), is barely enough to offset attrition caused by predation and the rigors of shorebird existence.

Breeding

Over the vast open spaces of the world, shorebirds breed in a wide variety of habitats. Some of these include Arctic tundra, beachfronts, marshes, grasslands, and tidal wetlands, although one species, Solitary Sandpiper, nests in boreal forests by using old songbird nests up to 40 feet

above the ground. These nests are mostly from American Robin, Canada Jay, Eastern Kingbird, Rusty Blackbird, and Cedar Waxwing. Of the world's 85 sandpiper species, only Solitary and Green Sandpiper from Eurasia routinely lay their eggs in tree nests instead of on the ground.

More than half of the 49 regular North American breeders use Arctic locations for breeding. A good number of these species travel great distances from breeding areas to wintering grounds, with some approaching 20,000 miles per year-round trip. These long journeys are filled with peril, and thus more reason for the importance of a high rate of reproductive success.

Since shorebirds do not give live birth to their young, they rely on a minimal nest scrape, often lined with vegetative material, to deposit their typical three or four eggs. One or both adults incubate the eggs until they hatch, with a time frame ranging from 17 days for Red-necked Phalarope to 31 days for Long-billed Curlew (smaller species have shorter incubation periods). This is a time filled with peril, as most shorebirds nest on the ground without much cover, and aerial predators can see the nests from the air. Ground predators like Arctic Fox and Wolverines can smell the nests while foraging for food.

Because of their tendency to breed in open spaces, most young shorebirds leave the nest with the ability to feed themselves from 1 to 24 hours after hatching. The biological term for this almost full independence of newly hatched young is precocial, which comes from the same Latin source as "precocious," meaning "ripened beforehand." Shorebird chicks fall into the category of Stage 2

Precocial, as they are not fully independent upon hatching but rely on one or both parents to protect and guide them and keep them warm or cool (warm-climate nesters) until they acquire a full set of feathers.

Since most North American shorebirds do not feed their young, with a few exceptions such as oystercatchers and American Woodcock, adults typically leave their chicks to fend for themselves as soon as they acquire a full complement of juvenile feathers, which typically takes less than two weeks. Only a handful of bird families enjoy this youthful independence, and they include ground-dwelling and ground-feeding fowl-like birds such as quail, turkeys, ducks, geese, swans, and domestic chickens. Other examples of precocial birds include Ostrich, pheasants, kiwis, and most shorebirds.

This is a very different scenario from most other bird families that are altricial, where the blind, naked, and helpless young hatch according to the timetable of egg laying, and where the chicks remain in the nest for weeks or months (such as Bald Eagle and Tundra Swan), fully dependent on the adults for food and protection during this time. Altricial comes from a Greek word meaning "wet nurse," which relates to the great amount of care the parents must administer their young for long periods before they become self-sufficient. Precocial eggs are proportionally larger than altricial ones, as they need a greater amount of nutrition to sustain the young for longer periods, and precocial chicks stay in the egg about twice as long as similar-sized altricial ones, since they need more time to develop in preparation for a quick departure from the nest.

◀ An incubating Black Turnstone is the picture of serenity and grace, and the tundra habitat surrounding it appears soft and inviting. A few weeks prior, this nest site was probably covered with snow. Incubating Black Turnstone is rarely depicted in photos, as the nest range for this species is far removed from roads or human habitation other than Native peoples. Y-K DELTA, ALASKA, LATE MAY

▶ "Oh, so cute!" Yes, but newly hatched American Golden-Plover chicks depend on their cryptic, tundra-calibrated patterning for protection from predators. The natal plumage of this young chick (about an hour old) with gold-spangled tones is one of the most attractive of all shorebirds. ALASKA, JUNE

Nesting

A number of interesting survival strategies occur during the egg laying, incubation, and brood rearing processes. After elaborate courtship displays that include unusual behaviors and vocalizations not seen away from breeding areas, the female typically lays three or four eggs, with most Arctic-breeding shorebirds laying four eggs unless it is a re-nest, when three eggs are often laid. Smaller shorebirds like Semipalmated Sandpiper can lay an egg a day, completing the egg laying process in a short four days. Larger shorebirds like Black-bellied Plover typically lay an egg every 36 hours or so and complete the laying of four eggs in six to seven days.

Eggs are strategically placed together in the center of the nest with the larger end out and smaller end pointing to the nest center to allow more exposure to the feathered belly of the incubating adult. This allows optimum warming or cooling of the eggs. Each time the adult returns to the nest, they reposition the eggs to maintain this effectiveness. We use the word nest here, but it is typically just a scrape on the ground, often lined with lichen, leaves, or other materials, such as Cotton Grass and feathers.

The nest period is filled with danger from predators, for both the incubating adult and the eggs. Ground mammals such as fox, raccoons, and wolverines are constantly searching for a meal during nesting season, and they use both sharp eyesight and a keen sense of smell to locate shorebird nests. Aerial predators such as gulls and jaegers patrol the skies in search of incubating shorebirds and swoop down to flush the adult off the nest before eating the eggs. A few Arctic-breeding shorebirds aggressively defend their nests from aerial predators, including Black and Ruddy Turnstones and Black-bellied Plover.

Cryptically feathered young are highly precocial, able to walk and forage for food mere hours after hatching, and requiring only protection from predators and the elements, and a nudge toward prime feeding areas by one or both parents. In many species, the protective umbrella spread

▲ A Ruddy Turnstone's high Arctic nest, often placed near driftwood, shows the strategic arrangement of eggs, with the larger end up to allow for greater surface area to be warmed by an adult's belly feathers during incubation. Each time an adult returns to its nest, it reconfigures the eggs to this placement. ALASKA, JUNE

▼ Black Turnstone, like its close relative Ruddy Turnstone, is a fierce protector of its Arctic nest and breeding territory. Flying fast and low across the tundra, turnstones explode upward toward the belly of the aerial predator, such as this Long-tailed Jaeger, which has no defense against this type of attack from below. Y-K DELTA, ALASKA, LATE MAY

over their young by adult shorebirds can assume heroic proportions. So determined is the territorial defense of Black-bellied Plover that Glaucous Gulls and jaegers give ground as soon as these vengeful black-and-silver missiles climb toward the transgressor.

Only a handful of bird families are as independent after hatching as shorebirds. Fledging (able to fly) in as little as 14 days, most juveniles migrate after the adults have departed. A staggered migration strategy reduces competition for food along migration routes, thus increasing foraging success at a time when birds need to lay down layers of fat to fuel their migration.

Because of this ever-present danger, shorebirds have worked out a few nest protection strategies to assist in their success. One is the use of nest distraction behavior where the non-incubating adult shows itself prominently away from the nest and acts as a decoy. After gaining the attention of the nearby potential predator, it moves away from the nest site and behaves as if it were returning to the nest, including sitting down on fake nests. This behavior works very well on inexperienced shorebird biologists searching for nests, who can't fathom that a creature with a brain the size of a pea or bean can outwit a highly educated biologist. It usually takes a full season of humbling

Black-bellied Plover is arguably the most fierce, aggressive shorebird protector of Arctic breeding territories. Flying swift and low across the tundra, it powers upward into the airspace of aerial predators like this Parasitic Jaeger with precision attacks, usually resulting in the hunter's rapid retreat. Other shorebirds nest near this species for secondary protection. This superb illustration was made by biologist Sophie Webb for a team t-shirt while working with Kevin for Troy Ecological Research Associates on Alaska's North Slope in 1994.

When its eggs are about to hatch, Black-bellied Plover performs animated nest distraction displays to draw the predator, or in this case biologist, away from the about-to-hatch chicks. In the body language of this bird: "You want those eggs? You've got to get past me, first." Note the spiky black wingpit feathers in this large plover species. ALASKA, JUNE

experiences before the biologist recognizes the distraction behavior of experienced adult shorebirds.

If this strategy does not work, the incubating shorebird sneaks off the nest and pretends to be injured, running away from the nest site in a zigzag motion with its wings dragging on the ground (see page 86, lower right photo). Predators often fall for this distraction, as they believe that an easy meal is imminent, and they usually chase the "injured" shorebird a good distance away from the actual nest. Another last-ditch behavior that occurs just before the eggs hatch involves an adult running at the predator with its wings out in a menacing fashion in hopes of scaring it away or distracting it from the nest. This is usually not successful, as most predators are savvy to this behavior, but Mountain Plover uses this scenario to successfully redirect the movements of cattle and Pronghorn Antelope that wander toward its nest.

The eggs of shorebirds are also cryptically patterned to blend in perfectly with the habitat where they are laid. Most have a spotted camouflage pattern that is very difficult to see, even at close range (see page 198, upper left photo), and casual human visitors and biologists need to be extra careful not to step on the eggs. Beach-nesting shorebirds lay eggs that are mostly pale with dark specks that resemble sand patterns, and these are also mostly invisible to the eye, even in bare nest scrapes in the sand.

◀ When the eggs of American Golden-Plover are about to hatch, this species performs a variety of nest distraction displays. The last line of defense is running at the predator with wings spread, hoping to distract them from the eggs. This male who charged at Kevin and displayed right at his feet seems to be saying "No, don't go over there, your meal is right here." It is truly noble defense against the equivalent of a 50-foot-tall opponent in human terms. ALASKA, JUNE

◀ Cryptic camouflage is the mainstay for survival of the eggs of American Oystercatcher. With adults often abandoning incubation duties to posture in aerial displays with neighboring oystercatchers, it is crucial that their eggs are virtually invisible to flying predators. NJ, MAY

Synchronous Hatching

After the incubation period, shorebird chicks from the same clutch usually hatch within a few hours of each other. This nest strategy is crucial to the survival rate of shorebird chicks, and to the species' overall long-term existence.

Once the first few eggs are laid, adults sit on them only intermittently to keep the eggs from freezing or overheating. This intermittence also prevents the viability of the incubation process from beginning. Adults sit on the eggs continuously only after a full clutch of eggs is laid, thus starting the incubation process for all the eggs at the same time. This results in the hatching of all the chicks within a few hours of each other and is known as *synchronous hatching*, crucial for shorebirds to maximize the number of chicks that survive.

If adults incubated full time after the first egg was laid, the first chick would leave the nest to forage long before the remaining eggs hatched, and the adults would instinctively attend to the first chick or two, leaving the other eggs exposed to the weather and predators, and without crucial adult care. At best, only two of the four chicks would hatch successfully and survive. With all the chicks hatching within hours of each other, the adults care for all of them, similar to a mother hen caring for all her chicks.

Synchronous hatching strategy guarantees that most or all the chicks receive focused care and brood rearing and allows many of them to reach fledgling stage. This complicated strategy for maximum nest success is not taught, but a result of imprinted information in shorebirds' brains.

To get out of their enclosed shell, shorebird chicks use a calcium protrusion on the upper bill tip (the egg tooth) to chisel away at the uniformly smooth, unbroken interior (see the photos below). Soon after hatching, this "egg tooth" falls off, and the chicks are defenseless for up to several hours as they stretch legs and flightless wings that have been cooped up inside a shell for three or more weeks. During this time, oxygenated blood begins to flow throughout their now movable bodies.

Brood Rearing

Shorebird chicks are up and feeding themselves within the first day, with some species such as Killdeer, Marbled Godwit, Long-billed Dowitcher, and Red-necked Phalarope (among others) leaving the nest as soon as their downy fluff dries. In many sandpiper species, the female departs as soon as or slightly after the eggs hatch, leaving the male to care for the young, including keeping them warm or cool until their juvenile feathers start to grow in (usually 7–10 days, depending on the species) as well as distracting potential predators. In some sandpiper species, such as Greater Yellowlegs, the female alone may tend to the young, and in a handful of others, both parents pitch in to care for the chicks.

Adults brood the young by giving a distinct call that alerts the young chicks that it is time to come under their wings and belly to rest or sleep, or for protection from the weather or predators. This is where the familiar saying "taking you under their wing" originated. With plovers,

▽ With planned precision timing, all 4 Black-bellied Plover eggs (*left*) are "pipping" (note the cracked openings) and about to hatch within hours of each other. Shorebirds start full incubation only after all eggs are laid (6–7-days for large plovers), thus enabling the viability of the incubation process to begin. Its purpose fulfilled, the calcium egg tooth on the bill in the right photo that allows the chick to chisel its way out of the smooth, contiguous egg wall will soon be shed, and all 4 eggs will hatch in a few-hour period rather than days apart. This amazing strategy affords all the chicks a much greater chance of survival in the harsh Arctic environment. This scene has rarely been witnessed or photographed. ALASKA, JUNE

Piping Plovers, like all shorebirds, care for their chicks by brooding them (tucking them under their body and wings). This male has excavated a shallow brood cavity in the sand and is calling his two-day-old chicks to come under his warm feathers. In beach-nesting shorebirds, this brood process may also serve to cool them in the hot summer sun. NJ, JUNE

oystercatchers, stilts, and avocets, both parents tend to the young after hatching.

After juvenile feathers grow in, the remaining adult sandpiper leaves the chicks to fend for themselves while they are still flightless (16–45 days, with smaller shorebirds fledging quicker). This is a perilous time for the visible young chicks, with danger around every corner on the ground and from the air. Gulls, jaegers, foxes, and other predators are prepared for this vulnerable period, and they work hard to find and eat the flightless young. In some shorebird species, such as the oystercatchers, both adults care for and feed the young for longer periods, up to several or more months.

Meanwhile the nonattending adults feed continuously in Arctic habitats for up to a month or so before starting their southbound migration. The chicks often gather in groups as they wait for the power of flight, after which they form feeding flocks before migrating a few weeks after the adults. These juveniles undertake amazing long journeys of up to 8,000 miles without adult guidance and arrive at the same general locations as the adults as a result of genetic information imprinted in their brains! Juveniles of a few shorebird species, such as Purple and Rock Sandpipers, migrate with adults.

Arctic-Breeding Shorebirds

As mentioned, more than half the regular breeding shorebirds in North America do so in Arctic or subarctic regions. These birds face incredible hardships during the breeding season, including often-frigid summer temperatures and a host of savvy predators.

These harsh environments initially appear unsuitable as breeding locations for long-distance shorebird migrants, but they provide some benefits that help with successful nesting. A few of these are the 24 hours of daylight for up to 70 days in summer that allow for constant food gathering, and the rapid propagation of insects and their larvae that shorebirds rely on to feed themselves and their young. Another is the lack of bushes, trees, or geographic features where mammalian predators can hide, thus enabling shorebirds to see danger coming at great distances.

After surviving a long return journey full of pitfalls, shorebirds work fast to squeeze in successful nesting during the short Arctic summer. They feed continuously while seeking a mate, after which a nest is constructed and eggs are laid. If a predator takes the eggs, Arctic shorebirds have only a short time to repeat the nest process (usually until late June), as accumulating snowfall and bad weather come as soon as mid-August in high Arctic regions.

In some shorebird groups, including the large plovers, the male builds the nest and adorns it with decorative vegetation. The female then either accepts or rejects the nest. If the nest is rejected, the male constructs additional nests until the female accepts one. Afterward, the female lays three or four eggs. Imagine the physical stress of forming and laying one egg every few days, with the four eggs weighing as much or more than the adult female!

Temperate-Breeding Shorebirds

Apart from Arctic, subarctic, and taiga breeders, other shorebirds nest mostly on beachfronts, grasslands, marshes, river and lake edges, woodlands, and high-elevation or

This photo shows Arctic tundra from the North Slope of Alaska, including moist graminoid tundra, which makes up about 80% of the 186,000 square miles of tundra above the Brooks Mountain Range. The raised areas in the foreground are drier high-centered polygons, prime nest habitat for large plovers. ALASKA, JUNE

A male Ruddy Turnstone is shown sitting on an open tundra nest on the North Slope of Alaska in June. This species is a fierce protector of its nest and breeding territory, and woe to any aerial predator that ventures into Ruddy's airspace.

A few-days-old American Oystercatcher chick plays dead near some seaweed after its parents went off to posture in flight with several pairs of neighboring oystercatchers. The chick's camouflaged plumage renders it virtually invisible from the air, and the lack of motion also protects it. This behavior is ingrained in the chick's DNA and needs no prompting from adults. NJ, MAY

alpine locations. Nest biology, strategy, and behaviors are mostly similar regardless of habitat and location, though slight differences have evolved in each habitat.

Beachfront nesting is similar to barren Arctic tundra breeding in that both locations lack any appreciable vegetative cover to retreat to for safety. In North America, three of the four small plovers (Piping, Snowy, and Wilson's) nest on open beaches or adjacent alkaline salt flats, and American and occasionally Black Oystercatchers nest on open beaches adjacent to oceans, bays, and the Gulf of Mexico. Eastern Willet is a large sandpiper that nests in grassy areas or marshes that border the Atlantic Ocean and tidal areas with brackish water, including the Gulf of Mexico, while its relative Western Willet breeds in grassy areas adjacent to or near western interior freshwater ponds or lakes.

Without appreciable cover, shorebirds that breed on open beaches protect their nests by performing distraction displays when danger approaches, including a "broken wing" display. Another protection for the eggs and chicks is their incredible camouflage that blends in perfectly with the habitat where they nest. These camouflaged eggs and chicks make it extremely difficult for aerial predators such as gulls and crows to see them, especially if the chicks are not moving. If an aerial or ground predator comes near, the adult gives an alarm call, which commands the chicks to play dead and stop moving. It is difficult even for ground predators to see the chicks crouched down on the sand if they are not moving, and they must rely on their keen sense of smell to locate the motionless chicks.

Marsh nesters often lay their eggs on small islands in ponds or marshes, which prevents smaller mammals from reaching their nests, and grassland breeders typically nest in taller grass areas (often near ponds), which provide excellent cover from both aerial and ground predators due to the uniformity of habitat over large areas.

Migration

Overview

Migration is the regular movement of creatures from one location to another for a variety of reasons. The most widely known type of migration occurs on a seasonal basis, where individuals arrive on the breeding grounds in spring and travel back to warmer climates at the end of their breeding cycle. Most shorebirds fall into this category.

Migration is energetically expensive and filled with peril, but it permits birds to avoid the privations of the northern winter and exploit seasonally abundant food resources in more temperate or tropical regions and the waxing resources of the extreme Southern Hemisphere as these regions enter the austral summer.

In days of yore, humans watched birds set their wings to the sky and disappear beyond the horizon and could only wonder what distant shores were knit by their wings. Today we simply marvel that these creatures of flesh and bone, feathers, and muscle, can vault entire hemispheres, relying solely on their powers of flight and genetic programing.

During migration, many birds, including geese, fly in echelons, or V-shaped configurations. Such formations allow birds to conserve energy by drafting off the wingtips of the bird ahead, reducing drag and conserving as much as 50% of the energetic cost of flying solo. Fighter aircraft fly in formation for much the same reason. Migratory stopovers, as well as wintering locations, are contingent on abundant food resources, so birds apportion themselves accordingly.

An intentional blur photo captures the dynamic motion of Red Knot and Ruddy Turnstones as these two long-distance migrants explode into flight in May at Delaware Bay, NJ. Both species vault hemispheres as they migrate from the high Arctic to South America and back again in spring.

▶ Dunlin, Short-billed Dowitchers, and Semipalmated Sandpipers are shown during spring migration, when all form large flocks. These three species often join together in tight flocks when taking to flight after sensing danger. NJ, MAY

Shorebird Migration

Most shorebirds migrate after breeding, whether short or long distances. Cold-climate breeders migrate south to access temperate or tropical locations, while others migrate from high-elevation breeding sites to lower elevations or coastal areas where the weather is not as severe in winter. Some shorebirds don't technically migrate at all but remain in the general vicinity of their breeding sites, or a relatively short distance away. Others travel from inland to coastal locations to access better food sources in winter.

As a group, shorebirds undertake some of the most spectacular long-distance migrations of any creature on the planet. These long distances are possible because of their relatively light weight due to hollow bone structure and the evolution of aerodynamically advanced feathered wing design in long-distance migrants. While flight patterns on breeding and wintering grounds appear graceful and almost effortless, long-distance migration requires direct powerful flight and a few physiological changes that are hard for us to comprehend.

Shorebird migration lasts a long time, unlike that of most other bird families. "Spring" migration begins in February for some extreme southern wintering species and ends in early June, while southbound "fall" migration starts in late June for some failed breeding adults and continues until mid-November, typically with juveniles. There are barely a few weeks in between spring and fall movements when shorebirds are not in migration mode.

Migration strategies vary greatly among shorebirds, with some species undertaking long, nonstop migrations while others make a few or numerous stops along the way to recharge their energy and replenish body fat reserves, thus accessing a variety of often-unfamiliar habitats during these journeys. Some of these stops comprise a few days, while others might last several weeks.

An example of this last scenario involves Buff-breasted Sandpiper, which leaves its wintering areas in southern South America in February and follows the north/south mountain ranges until reaching Texas and other south/central US locations in mid to late April. They fatten up in rice fields, wet meadows, moist lawns, and other agricultural areas for several weeks, after which they take up to a month to finish their protracted journey to high Arctic breeding grounds.

A good number of long-distance migrant shorebirds use a protracted migration like this to replace non-crucial feathers, such as head, body, and back feathers (scapulars), enabling them to molt partially into or out of breeding plumage while enjoying the benefits of food and rest during this energy-draining feather replacement. This strategy also allows them to dedicate all their time to setting up a territory and finding a mate upon arrival at their breeding grounds.

Shorebirds do not migrate in family groups like geese, swans, or cranes, except for some migratory oystercatchers where the young depend on their parents for food extraction for several months. In both spring and fall migrations, there is often a difference in timing between the movements of females and males, with males eager to get back to the breeding grounds in spring to set up territories before females arrive. After breeding, females of many species begin their southbound migration as soon as or slightly after the eggs hatch, often from late June to early July, leaving the male alone to care for the young.

What It Takes to Be a Flying Machine

Shorebirds are flying machines, and their physiological refinements are specifically designed to meet the extraordinary metabolic demands of flight. One adaptation is weight reduction, as the bones of most birds are hollow, and the feathers that encase them not only help to control their body temperature but also serve as airfoils (enabling lift or flight) at a weight cost of "light as a feather." And flight muscles of shorebirds are huge, constituting up to one-third of their total weight.

Shorebirds have also enhanced respiratory and circulatory systems, with highly efficient four-chambered hearts up to twice as large as those of mammals of comparable size, and lungs whose exchange of exhaust carbon dioxide with fresh oxygenated air approaches 100% with every breath cycle. In addition, the higher metabolism of birds facilitates the delivery of fuel (or metabolic material) to meet the energetic demands of flight muscles working at peak capacity. In shorebirds, these muscles as well as the heart and lungs increase in size prior to migration.

Feathers, those remarkable instruments of flight, are subject to wear and must be replaced each year by molt. To ensure that wings are operating at peak efficiency, many long-distance migrants wait until reaching the wintering grounds before replacing wing and tail feathers (the most crucial flight feathers). The very body shape of shorebirds is an aerodynamic, football-shaped marvel designed to reduce drag and propelled by scimitar-shaped wings with low camber arc and high aspect ratio to further reduce wind drag.

The fuel for long-distance migrants is stored fat, laid down in layers beneath the skin, with some infused through the muscles themselves. It is precisely the capacity for rapid weight gain (hyperphagy) that makes long-distance flight possible while using a minimum of food-rich staging areas. This stored fat, which may double the normal weight of some species prior to departure, offers twice the energy and water of protein or carbohydrates, with this water necessary to keep birds hydrated during long, non-stop flights. A further weight-saving measure is the shrinking of unnecessary internal organs preceding these long flights (including the liver, kidneys, digestive tract, and gonads) to reduce wing loading.

Despite these modifications, most shorebirds carry only enough fuel to fly several thousand miles before having to stop at critical staging areas to refuel. Flying conditions determine the rate of energy consumption, and a favorable tail wind reduces metabolic demand. This more efficient use of energy is why many southbound migrants to South America fly out over the Atlantic Ocean, trusting the east-blowing trade winds to assist them on the long journey. Tail winds do not necessarily speed a bird's journey, but they allow them to throttle back and reduce fuel consumption.

As a last resort, if shorebirds exhaust their fat reserves, they begin metabolizing muscle tissue to fuel their flight. Protein is not as energy packed as fat, and when the engine begins to burn itself, the end is near. Twice a year golden-plovers, godwits, Red Knot, White-rumped Sandpiper, and other long-distance shorebird migrants play out this drama in the skies over two hemispheres, with evolution siding with the birds and a capricious universe offering long odds.

Spring and Fall Migration

Spring migration is shorter and faster paced than its fall counterpart. Many of the numerous Arctic-breeding shorebird species travel in large flocks in spring and tend to bottleneck at strategic stopover sites during a short few-week period from mid-April to late May, depending on the location in North America and the species involved.

They are racing to get to Arctic breeding grounds within days or weeks of the snow melt to begin their swift nesting season of a few months when they must define a territory, find a mate, lay eggs, and brood their young. In the high Arctic, this season starts in early June and is over by mid-August, when snow once again covers the ground until the next spring.

This timetable is crucial to nest success, so the birds tend to move north in large flocks during a relatively short time frame. Temperate-breeding shorebirds also move during this same window to ensure a timely arrival on their breeding grounds, where they must define breeding territories and attract mates before too much competition arrives.

Fall shorebird migration is more protracted and relaxed, and occurs in waves, beginning in late June and continuing into November. The first shorebirds to head south are failed breeders with no young to raise, and this occurs from late June to early July. Next are half of each sandpiper pair, usually females, who abandon the nest area right after the eggs hatch, or shortly afterward, leaving the other adult to care for the flightless young until a full complement of juvenile feathers grows in. After forming loose flocks and feeding for a few weeks, these adults begin their southbound migration, usually in mid to late July for high Arctic breeders, and slightly earlier for lower Arctic and taiga breeders.

Semipalmated Sandpipers and Dunlin are tightly packed in this large spring migrant flight at Heislerville, NJ in May, an important stopover location. These extended stops allow shorebirds to feed constantly and pack on more weight and fat for the remaining long journey to Arctic breeding grounds.

A nonbreeding Short-billed Dowitcher makes a running start to flight at Bolivar Flats, Texas in September. The streamlined football-shaped body and long scimitar-shaped wings are some of the physical adaptations that allow shorebirds to more easily cut through the air during long migratory flights.

Some shorebirds like Dunlin form large migratory flocks in spring that pause at important feeding areas to fatten up for a long remaining journey. Spring migration is faster paced than fall, as it is crucial to arrive at breeding grounds in time to complete the nesting process. A few Semipalmated Sandpipers are mixed in this mostly Dunlin flock at Heislerville, NJ in May.

A low center of gravity and short sturdy legs allow Purple Sandpipers to exploit the surf zone of rocky coasts. These adventurous birds feed where other shorebirds fear to tread, in the rushing waters of the Atlantic Ocean. This 1st-spring bird in mostly breeding plumage was photographed in early May in NJ.

Feathered black wingpits are visible on this juvenile Black-bellied Plover as it lifts off an Atlantic Ocean beach in NJ on New Year's Eve. Some juvenile shorebirds migrate thousands of miles to wintering grounds without guidance from experienced adults, relying only on accurate imprinted genetic information in brains the size of peas and beans. Simply amazing!

A few exceptions include Purple and Rock Sandpipers and Dunlin. Purple and Rock Sandpipers remain on or near their breeding sites for a few months after nesting before migrating to Atlantic and Pacific rocky coastlines for the winter in October/November, while Dunlin remains on or near its Arctic or subarctic breeding sites until late August or September.

Arctic- and subarctic-breeding male and female plovers, who stay with their chicks until they are fully feathered, migrate south from mid-July onward, depending on the nest initiation date or a successful re-nest. These plovers may be present in s. Canada or n. US locations as migrants from mid-July to October, and into early winter for Black-bellied Plover.

Fall shorebird migrants also use stopover sites to fatten up during their long journeys, but most species are not concentrated in nearly the same numbers as in spring. They also spread their migration out over a period of months in late summer and fall compared with weeks in the spring, as there is no dire urgency to get to their wintering sites. Shorebirds also use fringe wetlands and feeding locations besides the main stopover sites as refueling stops. Without the rush to breed, the migratory pace to wintering areas is relatively leisurely compared with spring migration.

Temperate-breeding shorebirds, including beachfront, grassland, marsh, and prairie nesters, utilize a wide range of arrival and departure dates for migration, depending on climate and other environmental factors. Some species, particularly beach nesters, may remain in the general vicinity of their nest site year-round, or only migrate relatively short distances to access a slightly warmer climate or better food sources, or both.

While most of the super long-distance shorebird migrants involve high Arctic breeders, some temperate shorebird nesters also migrate deep into South America. Some of these species include Eastern and Western Willets, Solitary and Spotted Sandpipers, and Wilson's Phalarope. Wilson's visits several important stopover sites to fatten up before undertaking long-distance, nonstop flights to winter in high-elevation lakes in the southern Andes, as well as in the Patagonian lowlands and Tierra del Fuego.

Juvenile Shorebird Migration

Most juvenile shorebirds typically leave the breeding grounds a few weeks to a month after adults and travel to specific shared winter locations with no guidance from experienced adults. These accurate initial flights are possible due to imprinted genetic locational information in shorebird brains that are only the size of peas or beans, and this information is similar in precision to our present-day GPS navigational systems. We don't know how this information is transferred into action, and it is truly one of the great mysteries of the animal kingdom.

Arctic-breeding juvenile shorebirds usually arrive at most Canadian and US locations from late July to September, depending on a variety of factors, including date of nest initiation, latitude of nest site, and duration of incubation and fledging. Juveniles often remain at migratory stopover sites much later than adults, sometimes into October and occasionally early November, before continuing to their wintering locations

Migration Perils

Although more than 20 million shorebirds migrate through the United States to the Arctic each year, Arctic shorebird biologist Pete Myers and his colleagues captured the attention of the ornithological and conservation communities with their discovery that the long-term survival of even abundant species may be in jeopardy.

Their studies show that Sanderlings, Ruddy Turnstones, Red Knots, Dunlins, and White-rumped, Baird's, Stilt, Western, and Semipalmated Sandpipers form enormous concentrations at several key staging areas along their migration route. Each of these spots is critical for successful migration of these species, providing superabundant food resources that enable the birds to quickly replenish their energy reserves and continue on their journey.

In North America, six such sites support millions of shorebirds annually: Alaska's Copper River Delta; Washington's Gray's Harbor; the Bay of Fundy in e. Canada; Kansas's Cheyenne Bottoms and Quivera National Wildlife Refuges; and the beaches of Delaware Bay in New Jersey and Delaware. More than 80% of the entire North American population of some species may join ranks at any of these key locations, with virtually all 4 million Western Sandpipers staging at the Copper River Delta site in May.

Other vital locations of similar importance are also recognized throughout the Americas, and these critical staging areas underpin the entire migration system of New World shorebirds. As Myers points out, such enormous concentrations dependent on so few widely spaced locales

Great numbers are no bulwark against population declines. Whole populations of some shorebirds like these migrating Western Sandpipers may be concentrated in a few key staging areas for short periods, making them vulnerable to natural and man-made perturbations. More than 4 million Western Sandpipers stage in Cordova, Alaska each May on the way to their Arctic breeding grounds.

break the usual link between a species' abundance and its immunity to population crashes or extinction.

The series of critical stopover sites is typified by Delaware Bay. The arrival and departure of 500,000+ shorebirds within a span of three to four weeks is synchronized with the annual breeding cycle of the Bay's enormous population of horseshoe crabs, for it is the eggs of the crabs that supply the energy required by the birds to complete their spring journey to the Arctic.

Each evening, after daylong feasting on crab eggs, the birds move east to roost in tidal marshes on the outer beaches of the Atlantic coast. Coastal and wetlands development have forced the birds into ever smaller foraging and roosting sites as the number of suitable areas dwindle. On high-tide nights, tens of thousands of shorebirds may be packed into a few hundred yards of beach.

Fortunately, efforts are now under way to link the key staging sites connecting wintering and breeding areas into a system of sister reserves. Shorebird biologists, backed by the World Wildlife Fund U.S., the International Association of Fish and Wildlife Agencies, and the National Audubon Society are working toward establishment of these critical reserves throughout the Americas. Success hinges on persuading local, regional, and national governments that such a system is not only desirable but absolutely necessary to ensure the survival of migratory shorebirds. As a first step, in May 1986, the governors of New Jersey and Delaware mandated the lower estuary of Delaware Bay as a reserve for shorebird conservation.

Amazing Shorebird Migratory Journeys

While it is no secret that many shorebirds migrate long distances to their breeding and wintering grounds each year, the how and why of such movements remain a puzzle to humans. For example, how do shorebirds know exactly where they are going over thousands of miles, especially during extended travel over open oceans, and how do they know when to start these return migrations when they are in the mid-latitude tropics, where daylight hours are near constant? Could it be the angle of the sun that precipitates their northward journey in late winter? Another mystery is why they travel in large groups during these journeys. There is much that we don't understand about shorebird migration, and even if they could communicate with us, they might just answer "because it works."

Here are a few examples of long-distance movements involving three species of super shorebird migrants.

Bar-tailed Godwit—The Champion of Migration

After breeding in the Arctic tundra, North American Bar-tailed Godwits add up to 55% body weight by continuous eating of tiny clams for about a month in August on tidal flats in w. Alaska. Then they wait for predictable strong storms in late summer to get a tail wind that takes them about 1,000 miles south with little effort. Without these

◁ Tens of thousands of migrant shorebirds are packed into 20+ miles of narrow beach during spring migration on Delaware Bay, NJ. Red Knots, Ruddy Turnstones, and Sanderlings constitute a large portion of shorebirds at this critical migratory stopover location in May.

A female Bar-tailed Godwit on its Alaskan breeding grounds on the Y-K Delta in late May. After nesting, she undertakes the longest nonstop migration of any creature on the planet, flying more than 7,000 miles in 7–11 days from Alaska to New Zealand and Australia.

Hudsonian Godwits undertake remarkable long-distance migrations to and from their boreal forest/taiga breeding grounds in Canada and Alaska. A handful of journeys of several thousand miles each are coupled together during these epic flights, with a few important extended feeding stops mixed in.
ALASKA, MAY

storms, they would not be able to physically complete their amazing nonstop migratory journey of 7,000–9,000 miles to New Zealand over a period of 7–11 days. With this tail wind, they can fly as fast as 60 miles per hour!

This is tantamount to humans eating continuously for three or four weeks and doubling their body weight, and then running nonstop for 1,000 miles over seven days without the benefit of food, water, or sleep. It truly is a physical feat approaching miracle status, and well beyond human capability or comprehension.

A nagging question Kevin always pondered while working on the Alaskan Arctic tundra is, "why do shorebirds go so far when there is ample wintering habitat much closer to the breeding areas?" These questions remain unanswered 30 years later, although complete glaciation of much of North America except for Arctic tundra habitats north of the Brooks Range during the last Ice Age more than 10,000 years ago might have set a pattern that has not changed today.

To add to these amazing physical feats, Bar-tailed Godwit shrinks non-critical body organs as a weight-saving measure for this flight. Kidneys, liver, and intestines that are not used during the 7–11 day flights become lighter to facilitate the journey and represent another part of this miraculous adventure in which a combination of factors is necessary for the success of this migration.

Hudsonian Godwits—Kendall and Sig

Recent satellite transmitter data show that Hudsonian Godwit undertakes similar marathon nonstop migrations from Canada and Alaska to southern South America, and back again in spring. Hudsonian is a large shorebird that breeds in North American boreal forest habitats (about 67% of their population) as well as in subarctic tundra, and like Bar-tailed Godwit it undertakes some of the longest nonstop migrations in the world, with a good part of these flights occurring offshore over Atlantic and Pacific oceans.

The Canadian breeding population stages near Hudson and James bays in Canada after breeding and then flies out over the Atlantic Ocean in late summer to fall (juveniles) to travel mostly nonstop to South America, where they often encounter bad weather storms en route. The Alaskan breeding population, which is a separate subspecies, also makes similar long, nonstop migrations to southern Chile and back again in spring, with a few nonstop legs of several thousand miles each.

This excerpt from an article titled "Hudsonian Godwits Go Long" (Watts and Smith 2014) documents the fall migration of two Hudsonian Godwits fitted with satellite transmitters on the Mackenzie Delta in nw. Canada by Dr. Fletcher Smith, a shorebird biologist with the Center for Conservation Biology in the summer of 2013.

> *Kendall, an adult male, left the breeding grounds on 12 July and flew 1,529 miles to Churchill (Manitoba Province) in just over 2 days, staged for 2 weeks before moving down to Hudson Bay to stage for an additional 4 weeks. Sig, an adult female, left the breeding grounds on 10 July and flew to Churchill in just over 70 hours and staged for an incredible 3 months along Hudson and James Bays. Kendall left Hudson Bay (after 4 weeks) and made a dramatic 3,900-mile, nonstop, 5-day flight to the Orinoco River Basin in Venezuela, where he staged for 3 weeks in wetlands and agricultural fields. He then moved to the Amazon Basin to stage for an additional month, leaving Manaus, Brazil and flying to Santa Fe province Argentina. Kendall moved on to Buenos Aires and out to Samborombón Bay.*

Sig followed a similar pattern, flying 3,124 miles in 4.5 days from James Bay to coastal Colombia where she staged for more than 3 weeks in Los Flamencos Nature Sanctuary along the Caribbean Coast of Colombia. From here, she flew to Bolivia, staging for 23 days before flying to Santa Fe Province in Argentina. Both these birds have spent considerable time within agricultural landscapes containing high densities of ponds.

Red Knot—"Moonbird, B95"

Red Knot is an American Robin-sized shorebird whose annual migration is one of the longest of any creature on the planet. Flocks of Red Knot migrate along the Atlantic Flyway between Tierra del Fuego, Argentina, and their breeding grounds in the high Canadian Arctic, which encompasses about 20,000 miles a year! This subspecies of Red Knot (*Calidris canutus rufa*) usually makes this northbound epic journey with only a few stops, culminating at a major staging area in May along the Delaware Bayshore of New Jersey and Delaware.

Red Knots arrive famished and undernourished (sometimes weighing as little as 130 grams) along Delaware Bay and feed continuously on horseshoe crab eggs for about two to three weeks, when they strive to achieve a desired body weight of 200–225 grams, or just under half a pound. This extra fat reserve enables them to make the final nonstop flight of 2,875 miles to the Canadian Archipelago high Arctic region, where they breed. There are no adequate stopover feeding sites along this final leg of their journey, so if they don't achieve sufficient body fat reserves, they will drop into the vast boreal forests of Canada, where they search for life-sustaining riverbeds. These birds will typically perish or not continue to breeding areas.

While the movements of Red Knot are legendary, it took a very special individual to bring international attention to

◁ Red Knot is one of the champions of long-distance migration, flying up to 20,000 miles round trip each year from high Arctic breeding grounds to southern South American wintering locations. A few important stopover feeding locations are included in these flights, including Delaware Bay. NJ, MAY

◄ A breeding Red Knot feeds on horseshoe crab eggs washed ashore in the surf line of Delaware Bay in May.

▼ A brine fly's-eye view of impending trouble cast in the form of a meal-minded Wilson's Phalarope by Lloyd Spitalnik. This somewhat terrestrial member of the phalarope clan winters mostly in hypersaline lakes in the high Andes of South America, not in open oceans like its two family members.

NYC, LATE AUGUST

this story. In February 1995, Argentine biologist Patricia González put an orange band on the leg of a Red Knot in Río Grande, Tierra del Fuego, Argentina. The bird was at least two years old, and the band number was B95. This was the beginning a long saga that captured the attention and affection of millions of people worldwide and resulted in many articles and a book about this bird.

Why all the attention? The bird, a male Red Knot, was resighted a number of times over the next two decades along the Delaware Bay in New Jersey, and it was also spotted in May 2014 by Dr. Patricia González in the Canadian Arctic. It was recaptured at least three times (the last time in 2007) and sighted for the last time in May 2016 along the Delaware Bay, making the bird at least 21 years of age.

A little math determined that B95 had flown more than 400,000 miles over the course of those 21 years—roughly equivalent to flying to the moon and halfway back again. B95 was fondly given the name "Moonbird," and Phillip Hoose wrote an award-winning book about him titled *Moonbird: A Year on the Wind with the Great Survivor B95* (2012).

This publicity brought public attention to the perils of shorebirds and their migrations, and the importance of supporting conservation and research by shorebird biologists. With more than just facts and figures, the personification of B95 brought this struggle down to the human level of emotion and understanding, which is crucial to the future of shorebirds' survival and their support. There is a statue of B95 in Mispillion Harbor on Delaware Bay, and the city of Río Grande in Tierra del Fuego is said to have proclaimed B95 its "natural ambassador."

Shorebirds in Winter

Outside the breeding season, shorebirds are mostly social and coastally concentrated, where they find an abundance of invertebrate prey. Of course, there are exceptions, including those members of the grassland guild (Upland Sandpiper, Buff-breasted Sandpiper, and American Golden-Plover) that remain true to form and winter in the Pampas grasslands of Argentina and Uruguay, far from coastal areas, where they feed on an abundance of grasshoppers. Wilson's Phalarope is another outlier, wintering mostly in hypersaline lakes in the Andes, while Red and Red-necked Phalaropes mostly winter at sea off the coasts of the southern continents.

In fall and winter, most shorebirds are cryptically and often blandly plumaged, and many species join large mixed-species flocks that are ever vigilant for raptors that specialize in hunting shorebirds, most notably Peregrine Falcon and Merlin. Predator pressure also affects winter shorebird numbers and their distribution. At Bolinas Lagoon in California, hunting by Peregrines, Merlins, and Short-eared Owls is estimated to kill up to 20% of wintering Dunlin.

And while it is safe to say that many shorebirds winter in the Southern Hemisphere, there are quite a few that don't. Some of our southern coastal plovers (Snowy and Wilson's) remain for the most part in their natal area, gathering in small and often mixed flocks on coastal beaches. Jacana is essentially resident, relocating locally as water conditions dictate. Mountain Plover is a short-distance migrant, relocating from prairie breeding grounds to dry interior plains in California, the Southwest, Texas, and northern Mexico, where they seek out well-grazed grasslands and increasingly fallow agricultural fields caused by the loss of native grassland habitats. Some of the Beringia breeding Rock Sandpipers are

▲ Merlin is the main aerial predator of small- to medium-sized shorebirds, as this unfortunate Western Sandpiper in Cordova, Alaska can attest. Rapid flight and ability to change directions on a dime allow Merlin to pursue and catch retreating shorebirds, as does their ability to surprise roosting shorebirds by flying low and fast across open spaces. ALASKA, MAY

▼ Like an M. C. Escher illustration, a large flock of American Avocets at Bolivar Flats, Texas in April feeds in typical synchronous fashion after finding a concentration of small bait fish. Note the orderly layers of feeding birds with their back feathers (scapulars) spiked up in excitement. This is truly an amazing spectacle to observe.

essentially resident, relocating from tundra breeding areas to adjacent coastal locations, where they often gather in large flocks outside the breeding season.

Conversely, there are shorebirds whose latitudinal wintering range spans both hemispheres. Sanderlings winter coastally from New England and British Columbia south to the Straits of Magellan, and Surfbird winters coastally from Kodiak Island, Alaska, to the southern tip of South America. This extensive distribution affords a level of protection in the same way that a diverse portfolio imparts economic security.

But this does not explain why an individual Sanderling might elect to fly the additional 6,500 miles to Tierra del Fuego when there are Sanderling-calibrated sandy beaches and lots of wintering Sanderlings for company on Long Island, NY. If nothing else, the Sanderling's apparent indifference to migratory stress is a tribute to how superbly refined these long-distance migrants truly are. It is worth noting that among Sanderlings, there is no difference in the latitudinal wintering distribution of adults compared with juveniles, but many juvenile Sanderlings that winter in South America do not return to Arctic breeding grounds in their second year.

When not roosting or avoiding predators, much of a shorebird's daily nonbreeding activity pattern revolves around foraging, mostly at low or falling tide. Birds wintering in colder climates necessarily feed more heavily than those in temperate regions, as they burn more energy keeping warm. The doughty Rock Sandpiper winters mostly in coastal Alaska and forages on every low tide cycle, day or night. Even in areas rich in small bivalves,

Rock Sandpipers must spend at least half their daily time budget feeding to keep up with the high metabolic demands of wintering in coastal Alaska. This helps to explain why Sanderling and other *Calidris* sandpipers are willing to invest the energy they do in migration, the energetic equivalent of paying forward.

Wintering shorebird energetics are further complicated by the need to replace their body and flight feathers to attain breeding plumage, a transformation that typically starts in late winter, just prior to migration, and continues during migration, especially at important extended stopover sites. Feather replacement consumes lots of energy, but the biological payoff is the privilege of moving your genes forward.

It is precisely in winter that shorebirds reach their highest level of artistic expression. During the breeding season, they are spread out over thousands of miles, but in massed flocks in winter, we find shorebirds in their aggregate finest, foraging in massed waves and wheeling in synchronous clouds to foil hunting raptors. The ranks of Sanderlings playing tag with the waves impart an element of animation to the winter-scoured landscape, while the thick, crowded ranks of American Avocets mantling the Bolivar Flats recall the works of the Dutch graphic artist Maurits Cornelis Escher.

When the tide turns, sending waves of Dunlin in search of higher ground, the air fairly sizzles with their rapid wing beats, and onlookers marvel that the tiny birds can pack so much enchantment into their twice-daily commute. We are not accustomed to this level of intimacy with birds, but shorebirds, because of their commanding attributes, can

A flock of hardy Dunlin sit out a frigid NJ winter's day on the Atlantic coast in December when the temperature was 10 °F with a wind chill below zero. Kevin was able to walk through this pack without flushing a single bird. They just shuffled aside and quickly put their bills back under their warm back feathers, hoping Kevin would go away without harming them.

As the hardiest of all shorebirds, these Rock Sandpipers are spending the winter near breeding sites in w. Alaska, often in frigid, icy conditions where food is hard to come by and the weather is less than hospitable. More Rock Sandpipers than rocks in this photo by Brian Guzzetti in February.

afford to be both confiding and engaging. We repay their confidence with admiration, wonder and verse.

SANDPIPERS
Carl Sandburg

Sandland where the saltwater kills the sweet potatoes.

Homes for sandpipers—the script of their feet is on the sea shingles—they write in the morning, it is gone at noon—they write at noon, it is gone at night.

Pity the land, the sea, the ten-mile flats, pity anything but the sandpiper's wire legs and feet.

Shorebird Mortality and Declining Fortunes

With their somewhat low reproductive capacity, high natal mortality, and relatively short life span, becoming a breeding-aged shorebird might well constitute one of the animal world's greatest accomplishments. Literally living on the edge, shorebirds are exceedingly vulnerable to perturbations within their life cycle. Some shorebird populations have suffered an alarming 70% decline since the first book titled *The Shorebirds of North America* was published

in 1967 (Stout et al. 1967). The vast numbers that shorebirds have long relied on to sustain them no longer seems the bulwark it once was against population declines.

In some portions of the Arctic and subarctic, dramatically increased Snow Goose numbers have denuded great swaths of vegetative tundra that once supported breeding shorebirds, especially in the taiga-tundra ecotones of Hudson and James bays in Canada. Reduced snowpack has also caused regional reductions in lemming numbers, resulting in higher predation on nesting shorebirds by foxes, jaegers, and gulls that find shorebird eggs and nestlings an alternative to lemmings.

In migration, shorebirds face challenges fostered by humans, whose coastal development and efforts to combat sea level rise diminish shorebird habitat. Shorebirds also face human hunting pressure in protein-starved portions of their migratory and nonbreeding ranges, especially in South America and China. Changing agricultural practices and direct competition for food resources also levy a toll on the birds. A prime example is the overharvest of horseshoe crabs in Delaware Bay, depleting the eggs on which tens of thousands of migratory shorebirds depend to fuel up for the final leg of their journey to the Arctic. Indeed, the authors witnessed this harvest and worked to halt the decline.

Piping Plover, whose coastal subspecies ekes out a tenuous existence on the Atlantic shoreline, has disappeared from some beaches because increased human beachgoer traffic disrupts the nesting cycle and prevents chicks from

▲ A few-days-old Piping Plover chick has to maneuver around beachgoers to reach its feeding area in the Atlantic surf zone. Sharing ocean beachfronts with humans in summer, these cute, tiny puffballs with long legs are virtually unnoticed by their human neighbors, but feeding is often interrupted by the large number of beach enthusiasts. NJ, MAY

▶ A dapper male Piping Plover of the subspecies *C. m. circumcinctus* is part of an inland population of this vulnerable species that has increased over the last few decades, which is good news for this federally Threatened species. This increase will likely allow Piping Plover to survive the devastating effects of sea level rise, which threatens to extirpate the coastal subspecies in future years. FLORIDA, APRIL

reaching the water's edge to feed. The result is that nest productivity has fallen below the threshold needed to maintain local populations in a number of places. Their nest success has also declined as a result of climate change, with rising sea levels, higher tides, and summer storms washing out nests that would have been safely situated 40 years ago. These negative factors affect only the Atlantic coastal breeding population, with the inland subspecies only one of three shorebird species that have noticeably increased in numbers over the last several decades.

Year by year, challenge by compounded challenge, it is becoming harder and harder to become an adult shorebird,

and whole populations are faltering in some species. For now, the Arctic air still sizzles with the tinkling flight song of White-rumped Sandpiper, the whistled yelp of Golden-Plover, and the rhythmic hoot of Pectoral Sandpiper, yet ours may be the last generation to be enriched by these auditory treasures.

But this book is a tribute to shorebirds, not a eulogy. Shorebirds have endured comet strikes that killed off their dinosaur kin and market gunning that greatly thinned their ranks in the last half of the 19th and beginning of the 20th centuries. Climate change is just the most recent challenge faced by this bird group.

45

THE TRAGEDY AT DELAWARE BAY (ALSO HOPE FOR RECOVERY)

The earliest reference to the annual gathering of crabs and shorebirds on the beaches of Delaware Bay harks back to the early 1800s, including the account by ornithologist Alexander Wilson of the Ruddy Turnstone, one of the four principal shorebird species dependent on the annual tribute of horseshoe crab eggs to fuel their migration and breeding success. The other three species are Red Knot, Sanderling, and Semipalmated Sandpiper. Said Wilson of the density of mating crabs in his account of the bird that he called the "horse foot snipe" (Ruddy Turnstone): "A person could walk ten miles east from the mouth of the Maurice River on the backs of crabs and that stroller's feet would never touch the sand."

Such was the concentration of crabs in the early 1980s, before the ranks of crabs were decimated by the greed of "crabbers." While Wilson did not specifically link Red Knot to the eggs of the horseshoe crab, his name for Ruddy Turnstone ("horse foot snipe"), which forages generally higher in the bay than Red Knot, clearly indicates a historic link between the hosts of birds and crabs going back more than two centuries.

Nearly 90% of this crucial migratory food source was wiped out in the 1990s and early 2000s by uncaring opportunists who overharvested horseshoe crabs with no regard for sustainable yields or the birds that relied on their eggs for survival. At this time, there were no NJ state or federal regulations to limit the crab harvest, and excessive numbers were taken. Another major contributing factor to the crab's rapid decline was the bottom dragging of Delaware Bay, which not only yielded a large number of crabs but also disturbed the fragile ecosystem of these fertile spawning grounds. Much of the harvest resulted in crabs that were chopped up and used as bait for a local eel prized in Japan as a high-priced delicacy, but not sold or desired in the United States.

▼ A unique crab's-eye view from the Delaware Bay shoreline in May shows a mating frenzy with larger female horseshoe crabs attracting smaller, eager males. These crabs are the sole catalyst of the entire Delaware Bay shorebird phenomenon in spring.

▶ Ornithologist Alexander Wilson wrote in the early 1800s that the number of mating crabs on the Delaware Bayshore was so prolific that a person strolling could walk several miles without touching the sand. Numbers are greatly reduced today, but this small microcosm captures the phenomenon described by Wilson.

◤ As a large female horseshoe crab lays eggs just under the sand on the shores of Delaware Bay in May, nine smaller males deposit their sperm in the vicinity of the eggs to enable fertilization to take place. Note the tiny size of the lipid-rich crab eggs compared with a penny in the right photo.

This conflict over resources caused a great deal of consternation between environmentalists and the Delaware Baymen, who viewed the eventual protective regulations for horseshoe crab harvest as an infringement on their right to earn a living. At 25 cents per crab in the 1980s, only a handful of local Baymen were harvesting crabs, so the impact to the crab population was negligible at that time.

But when the price rose to $1.75 per crab in the mid-1990s because of increased demand for a rapidly shrinking resource, a good number of out-of-state businesses filled large tractor trailers with burlap bags of horseshoe crabs without concern for the decimation of their numbers. Local opportunists also joined the handful of Baymen and contributed to the overharvesting in the 1990s. With a

paltry license fee of $50 per season and no bag limits, the crab harvesters wiped out a large percentage of the crab population (fide Kevin Karlson).

Further exacerbating the problem was the purposeful selective harvest of large adult female crabs, which fetched a higher price on the market. Horseshoe crabs take about ten years to reach sexual maturity, and when you selectively harvest sexually mature female crabs, the entire population suffers in a short period of time. The impact on Red Knot numbers was catastrophic, declining from more than 100,000 birds on Delaware Bay in the early 1990s to a low of about 12,000 birds in the mid-2000s. This was a rallying cry for birdwatchers and environmentalists, and the conjoined voices of many supporters of Red Knots

 In the late 1980s and 1990s, high demand for horseshoe crabs for bait resulted in unregulated, indiscriminate harvesting by Baymen and local residents, as shown by the pickup truck with crabs crawling out onto the roadway, only to be crushed by vehicles. Only large, mature female crabs were taken for the higher pay, which devastated the population, as horseshoe crabs take about 10 years to reach sexual maturity. The population of crabs declined more than 90% in just over a decade, resulting in long-overdue regulations on harvesting these living relatives of arthropod trilobites from 400+ million years ago. NJ, MAY

and horseshoe crabs resulted in protective regulations for harvesting crabs enacted by the State of NJ in the late 1990s and mid-2000s.

After a decade or so, Red Knot numbers for the subspecies *C. c. rufa* rebounded to a high of about 40,429 birds, based on a comprehensive study undertaken on the Atlantic Coast on May 23–25, 2012, which found 25,548 knots in Delaware Bay; 1,500 in salt marshes around Stone Harbor, New Jersey; 8,621 in Maryland and Virginia; and 4,850 in North Carolina south to Florida. Add in the 2,000 knots counted in winter 2012 in the nw. Gulf of Mexico (D. Newstead, unpub. data), and the *C. c. rufa* population in 2012 was about 42,000 (Andres et al. 2012; Amanda Dey, pers. comm.). This higher number was despite a paltry record low of 9,850 wintering knots counted during aerial

surveys in southern South America in 2011, but this number rebounded to 13,000 birds counted in January 2012, with ground observations noting the presence of numerous juveniles reflecting a successful breeding season in 2011.

Larry Niles, from the Conserve Wildlife Foundation of NJ and Wildlife Restoration Partnerships, and Amanda Dey, from the NJ Dept. of Environmental Protection, Endangered and Nongame Species Program, head a team of international scientists and researchers who study, band, and count Red Knots and other shorebirds on Delaware Bay. This research began in the early 1990s and continues to the present day. They use the findings of their research to establish reliable data for the protection and ongoing monitoring of this bird that is living literally on the edge of survival, and to educate the public about the ongoing crisis

Groups of Red Knots, with a few Sanderlings, feed feverishly in a horseshoe crab egg concentration hole on a Delaware Bay beach in May. Note the worn orange leg band from Argentina on one of the knots, which Kevin hopes is "Moonbird—B95." This photo was taken in 2005, nine years after Moonbird was banded in 1996 by Kevin's friend Dr. Patricia González in Rio Grande, Argentina. Hope springs eternal, since the band is worn, and the numbers are not visible.

A shorebird banding operation is shown in this photo from May 1998 on Delaware Bay, NJ. Biologists gather important information, including arrival and recapture weights that indicate whether birds are physically prepared for the several-thousand-mile flight to high Arctic breeding grounds. Dedicated shorebird researchers like these are responsible for the Red Knot recovery in progress today.

of a species in peril. They also secured federal grant money after Hurricane Sandy for habitat restoration of parts of the Delaware Bay shoreline that are critical feeding areas for the birds that use this resource.

The ecological disaster that transpired in Delaware Bay has still not fully recovered after 25 years of protective management of the crab harvest, but recent trends indicate a possible recovery in progress. During the winter of 2022, numbers of Red Knots counted at key sites in southern South America were the highest since 2012, with 14,521 birds tallied.

In retrospect, however, it should never have happened. The spring concentrations of birds and crabs were among the ecological wonders of the world, and they were widely known to birdwatchers and ornithologists, well-studied and documented, and supported a thriving tourist industry. Two state and one federal agency were charged with protecting

the birds, and multiple conservation organizations vigorously decried the harvest.

New Jersey's Governor Christie Todd Whitman favored regulation in 1997, and the state even passed a few laws to protect the crabs, but local state legislators representing (they believed) the wishes of their constituents stymied efforts to fully halt the harvest. It is not the first time in our history that the needs of birds have gone head-to-head against localized self-interest groups and lost crucial battles, but regulations passed by Governor Corzine in 2004 and 2006 have further restricted the crab harvest, and a full moratorium on harvesting crabs for bait went into effect in 2008. Sadly, the progress toward recovery in the Delaware Bay is the exception, not the norm, which does not bode well for shorebirds facing increased challenges because of changing land and water use practices, here and elsewhere.

With increasing numbers evident today, a group of Red Knots and Ruddy Turnstones lift up from a Delaware Bay island in May, NJ. "Build it and they will come," a misquoted line from the movie *Field of Dreams*, is appropriate on the Delaware Bayshore after Larry Niles and associates secured relief money from the federal government after Hurricane Sandy in 2012 to replenish lost sand on the NJ Bayshore, with local businesses and government helping to truck the sand into place. Huzzah!

Market Gunning: "Sins of our Fathers

A Short Historical Essay

PETE DUNNE

It is difficult for those of us living in this conservation-minded age to contemplate a time when systematic slaughter of wildlife was not just sanctioned, but near universally accepted. We find it incomprehensible that members of our species would exterminate birds to the point where their very populations became threatened. But this is precisely what happened between 1800 and 1918, the years regarded as the "market gunning era." What's more, it all evolved quite naturally.

Today we brand those market gunners "game hogs" and "game bootleggers," but in their time, they were just ordinary tradesmen and valuable members of the community plying a trade that predates agriculture and animal husbandry. It was ingenuity and market demand that took a cottage industry and turned it into an industrial-scale business of slaughter.

Imagine the astonishment of those first Europeans to set foot on our shores. Immigrants fleeing a land of nutritional impoverishment whose protein needs were drawn mostly from the sea by the 1600s, finding in the New World a veritable Garden of Eden replete with innumerable quantities of "bustards" (turkey) and "fowl" (geese and ducks).

▽ When the early settlers first arrived in North America, the abundance of Wild Turkeys were a welcome sight, and many a meal was prepared for their families with this bounty. However, by 1830, John J. Audubon noted that overharvesting had resulted in the birds being hard to find.

There was certainly a seasonality to the bounty, when waterfowl blanketed the bays in winter and shorebirds made marshes vibrate with motion in spring and summer. There seemed no limit to the numbers, and no regulations governed the harvest of such bounty. It is likely that in no other corner of this planet was there a greater abundance of feathered game than was found in North America before the arrival of Europeans to these shores.

Villages sprang up, and men with primitive fowling pieces went out to harvest game to feed their families, just as native people were doing. A skilled hunter might bring home more game than he could use, so he shared his spoils with his neighbors, or perhaps he traded a goose or a brace of curlews to a neighboring farmer for a commensurate measure of corn or barley. In favorable times, he negotiated an exchange with the local innkeeper for a pint or two. There were no bag limits beyond a man's skill with a firearm.

From this low-key subsistence hunting and barter system, market gunning evolved. Villages became towns, and some towns became centers of commerce, which were later engines of the industrial revolution. Year after year, more and more Europeans flocked to the New World, especially Germans, Swedes, Irish, and Poles. By 1850, more than 23 million people were US residents (up from 2.5 million in 1776), and many of these new arrivals moved into the industrial centers of Boston, New York, Philadelphia, Baltimore, Charleston, New Orleans, and Chicago, where they found employment.

These city dwellers were deprived of access to the natural supermarket that had sustained earlier immigrants, so they were reliant on city markets to provide nourishment. What they craved in that age before Tyson chickens and Hormel hams was an inexpensive source of protein, a need that was served for many years by the slaughter of wild Passenger Pigeons and waterfowl. But after the demise of the Passenger Pigeon in the late 1800s and the gradual depletion of waterfowl, the market and gunners came to accept shorebirds as the ideal substitute. Large, abundant, and highly appetizing, the ranks of shorebirds were not only abundant but also obliging enough to ferry themselves to coastal marshes close to population centers. From the gunner's standpoint, shorebirds were close to ideal, as they flew in tight flocks, decoyed well, and were easily killed by a blast of fine bird shot.

An accomplished gunner could fill barrels with birds and reap a profit. In the days of plenty (1800–1880), only the larger shorebird species were gunned, with the smaller sandpipers not deemed worth the shot and powder. By the late 1880s, with Eskimo Curlew all but eliminated and Long-billed Curlews along the Atlantic Seaboard a distant memory, the guns turned upon the tiny "sand snipe." Gunners in Cape May would lie in wait for flocks of Least and Semipalmated Sandpipers and mow them down by the "hundreds" (Stone 1937).

Edward Howe Forbush, a noted Massachusetts ornithologist, writing in 1912, puts numbers to the slaughter: "116 yellowlegs killed with a single shot; 127 Red-breasted Snipe (dowitchers) killed in three shots" (John J. Audubon in Forbush 1912); 340 Wilson's Snipe killed in a single day on the Sangamon River in Illinois by two gunners. But these numbers pale compared with Audubon's account of 48,000

▶ Shorebirds come in a range of sizes, from the largest Long-billed Curlew to the smaller "peeps," like this Western Sandpiper. Long-billed Curlews were a favorite target for market gunners in the 1800s, when this species was eliminated from the Atlantic Coast.

(American) Golden-Plovers annihilated by a battery of gunners on March 11, 1821, near Lake Saint John, Louisiana (Audubon 1827). There is no reason to question the accuracy of Audubon's tabulation, nor any reason to believe that this was an isolated event. The slaughter was well planned by gunners who knew the habits of the birds through past experience. The sad diminishment of the ranks of American Golden-Plovers attests to the systematic slaughter.

Initially, the market gunning industry was held in check by a reliance on slow muzzle-loading weapons, but as weapons technology improved, so did the rate and deadliness of fire. The slaughter increased in measure. By 1855, the double-barreled breach-loading shotgun was in common use. Where a man armed with a muzzle-loading weapon might need minutes to reload, now with factory-produced shells, it could be done in seconds, just in time to fire a second charge at birds returning to the cries of fallen flock mates.

Bigger, boat-mounted weapons called Punt guns reaped a terrible toll on massed birds. Also employed was a diabolical practice known as fire-lighting, which involved two or more men. One brandished a mesmerizing kerosene lantern, and the other, the harvester, wore a gunny sack slung over his shoulder. With no firearm, he would sneak up on the bedazzled birds from behind, grab them by hand, and after administering a judicious bite to the neck, toss the limp form into the sack. It was primitive but very cost effective.

The other technological revolution that accelerated the slaughter was the railroad, whose capillary network of rail lines linked the killing fields of the prairies to market centers in the East. The spring shooting of breeding adult shorebirds was particularly impactful on bird populations, and in this age before refrigeration, many barrels of birds rotted before reaching markets.

Forbush (1912) offers an account of two Boston firms, which in 1890 received 40 barrels "closely packed" with

◀ One of the most attractive of all shorebirds, a male American Golden-Plover protests a disturbance near its tundra nest in Alaska by performing a distraction display. It is hard to believe that this species has come back from the indiscriminate slaughter by market and sport gunners in the 1800s, with 48,000 killed in a single day in March 1821 in Louisiana by a battery of gunners.

Buff-breasted Sandpipers numbered almost a million strong in the early 1800s, but were almost extirpated when market and sport gunners in the Midwest slaughtered them without mercy during their northbound migration for food and just for the sake of shooting them for "fun." Today their numbers are hovering around 56,000, although they remain a species of concern in North America (Andres et al. 2012).

Eskimo Curlews, Golden-Plovers, and "Upland Plovers" (now Sandpiper) from Nebraska, Missouri, and Texas. He goes on to lament that Golden-Plover had "almost disappeared" from New England, falling off 90% in 15 years.

While declines in some species were noted as early as the mid-1800s, and calls for restraint and an end to spring shooting came from some quarters, regulation of an industry so steeped in tradition and fueled by the myth of America's inexhaustible resources was slow in coming.

It wasn't until the passage of the Weeks-McLean Migratory Bird Act in 1913 that spring and night shooting were prohibited. But blanket protection did not occur until 1918 with the signing of the International Migratory Bird Treaty between the United States and Canada, which stated that all shorebirds except snipe and woodcock were given nongame status. The legislation was not in time to save the Eskimo Curlew, but it likely did save the Golden-Plover, Upland Sandpiper, and Buff-breasted Sandpiper from annihilation.

If market gunning brought out the worst in us, successive efforts to redress our crimes against nature reflect our best intentions to be good stewards of this planet.

These efforts began with the Weeks-McLean Migratory Bird Act of 1913; the Migratory Bird Treaty of 1918, affording protection to most migratory birds; the signing of the Duck Stamp Act in 1934, directing revenue generated by the sale of stamps to the migratory bird conservation fund; and establishment of the U.S. Fish and Wildlife Service in 1940, an agency overseeing migratory birds and the procurement and enhancement of land for wildlife. Today this federal agency manages 560 refuges, covering 150 million acres of prime (mostly wetland) habitat. In 1947, the Canadian Government followed suit with a Canadian Wildlife Service.

Other nongovernmental institutions working toward the betterment of shorebirds and the habitat they require include The Western Hemisphere Shorebird Reserve Network (WHSRN), The National Audubon Society, Ducks Unlimited, American Bird Conservancy, Manomet Bird Observatory, The US Shorebird Conservation Partnership, and Partners in Flight.

We cannot bring back the Eskimo Curlew, but we should be able to prevent another species from joining the curlew in oblivion.

"The beauty and genius of a work of art may be reconceived, though its first material expression be destroyed; a vanished harmony may yet inspire the composer, but when the last individual of a race of things breaths no more, another heaven and earth must pass before such a one can be again."

William Beebe, ornithologist

SECTION 2
Shorebird Species Profiles

North American shorebirds encompass six or more family groups as different from each other as warblers are from woodpeckers and thrushes are from jays. Things they have in common are an affinity for open, often semiaquatic habitats and a dedicated focus on invertebrate prey. All shorebirds are carnivores, and it is precisely the techniques they use to capture prey, in conjunction with physical adaptations, that facilitate these strategies. This permits plovers and sandpipers to forage in close proximity and not directly compete. Plovers and sandpipers account for most of the planet's shorebirds, but both confront the challenges of being shorebirds in markedly different ways.

A male Buff-breasted Sandpiper performs a single-wing courtship display at his mating post in a lek on the North Slope of Alaska in June. This display is done from a raised portion of tundra, allowing the intricate pattern of his white underwing to be seen by prospective females from a great distance.

Stilts and Avocets (Family Recurvirostridae)

These elegant, pied water striders are near cosmopolitan in their distribution, and they use their long legs, necks, and bills to exploit the invertebrate riches of shallow wetlands beyond the functional reach of less elongated shorebird species. The rapier-fine bills are straight and needle-like on stilts, and gracefully upturned in avocets, especially females.

Fresh, tidal, and hypersaline bodies of water are all acceptable habitats for these birds, particularly avocets, as long as they include an abundance of mostly invertebrate prey in the water column. The planet's seven Recurvirostridae species are closely related to the oystercatchers but could not be more different in terms of foraging techniques. Gone are the hardened chisel-like bills of oystercatchers, replaced by long, thin, flexible-tipped bills that are swished or stabbed into the water column and adjacent mud for aquatic invertebrates and small vertebrates (most notably small fish).

Avocets feed primarily by swishing their bills through the water column, a maneuver called "scything." They seine small food particles through their partially opened bills, which are then trapped in a complex system of lamellae and later whisked into the mouth by the avocet's fleshy tongue. Avocets are also visual hunters and may swim to feed in deeper water in pursuit of prey by using their partially webbed feet. Stilts consume mostly insects, crustaceans, amphibians, snails, and flies that are targeted mostly visually, although they can and do sweep their bills like avocets, snatching small fish from the water column.

Members of the Recurvirostridae family are among the most social of shorebirds, often nesting communally in small, loose groups. In winter, avocets gather in dense concentrations on coastal flats teeming with food, or at inland fresh or saltwater wetlands that don't freeze, most notably the Great Salt Lake. Stilts and avocets typically breed in shallow wetlands when local conditions are favorable to nest success. If conditions are not favorable in usual nest locations, birds may delay breeding or move to more suitable sites that provide reliable water levels, an abundance of prey, or reduced predation.

Both stilts and avocets nest in scrapes on the ground, often on small mud islands and typically in places with scant vegetation. These scrapes are often adorned with sticks, bits of bone, and shells (see page 63, left photo). A typical clutch is 3 or 4 eggs, and nonstop incubation commonly begins with the third or last egg. After hatching, downy young are able to run and feed themselves within a few hours.

Breeding success is variable from year to year, given the mercurial nature of shallow wetlands. Dropping water

▼ While generally tolerant of each other, it is apparent that a Black-necked Stilt wandered too close to an American Avocet's nest site. American Avocets are dominant where these two species share similar habitats. SALT LAKE CITY, UTAH, MAY

levels allow predators easier access to the nest site if a dry season occurs, and one good breeding season may be followed by several years of low productivity. Black-necked Stilts enjoy exceptionally long lives if conditions are favorable, with a possible life span of 19 or more years.

While the populations of American Avocets and Black-necked Stilts appear stable, the shallow wetlands that sustain them are vulnerable to modification, draining, drought, reduced water flow and pollution. Both these species have benefited greatly by the National Wildlife Refuge system, whose managed wetlands are ideal habitats for breeding and wintering stilts, avocets, and many other shorebirds. Buy Duck Stamps, which help to support the NWR system!

▶ Black-necked Stilt
Himantopus mexicanus

BIOMETRICS 14–15½ inches long; wingspan: 29–32½ inches; weight: 136–220 grams (4.8–7.8 ounces)
STRUCTURE Slender body with pointed rear end; extremely long legs; long, slim neck; long, needle-thin bill
STATUS Common in the West and South; scarce in Northeast; population appears stable or slightly increasing

This formally attired marsh bird strides through standing water on exquisitely long, red legs. Nicknamed Lawyerbird because of its crisp blackish upperparts and gleaming white underparts, and perhaps because of its verbosity,

▶ A dapper male Black-necked Stilt with pied plumage sports long legs that would make Betty Grable envious. Males have bluish-black backs in breeding season and long bubble-gum pink legs. TEXAS, APRIL

▽ After copulation, Black-necked Stilts walk side-by-side with affection and grace (*left*) as they celebrate the deed that was just done. Quiet, freshwater pools are favorite habitats for this species, including this pool in the oil fields near High Island, Texas (*right*). TEXAS, APRIL

△ It's rare to find so many quiescent stilts. A new arrival to the group (bird on the left) typically triggers a chorus of yaps from the high-strung "marsh poodles." TEXAS, APRIL

it is also somewhat irreverently called "marsh poodle." This is a clear reference to the bird's incessant yipping calls that it gives when excited, when danger is perceived, or just because a new member of the clan has joined the main group. Wherever they occur, Black-necked Stilts are among the most conspicuous and readily identifiable of all shorebirds.

Black-necked Stilts are high-strung birds and often react to any changes in their social grouping with aggressive posturing and loud "yipping" calls. Their breeding displays and behaviors are some of the most sensitive of any bird group, as evidenced by their post-copulation strutting where the male wraps his bill around the female's as they strut elegantly together while celebrating the special act that just occurred.

Stilts and avocets often share the same habitat, but stilts are generally less communal, although adults will congregate to add their voices to antipredator displays. While Black-necked Stilts are typically seen in small to medium groups of 5–30 birds in winter, large concentrations of up to 100 birds are possible. Stilts are also partial to freshwater wetlands with emergent vegetation, while avocets prefer open, non-vegetated mudflats with shallow water.

Black-necked Stilts have a widely spaced and disjunctive breeding range in the United States and Mexico south to southern South America, as well as the West Indies and Caribbean islands. In North America, the breeding range includes the Great Basin north to Oregon, Alberta, and Saskatchewan, and southern portions of the high prairies.

◁ Elegance in motion. The folks in Creation's R&D Department took the rest of the day off after rolling out the Black-necked Stilt line. A flock of these attractive birds always garners the attention of nearby birdwatchers.
TEXAS, APRIL

A female Black-necked Stilt with brown back (*left*) is a picture of elegance and symmetry in flight. A juvenile stilt (*right*) lacks the bold blackish plumage of adults and has a brownish wash to the back and head. TEXAS, APRIL

California and coastal Texas have a good number of year-round resident populations, and migratory populations occur from Delaware south to Florida and the upper Gulf Coast. Interior breeders other than those in California are migratory, and many southern breeders move only short distances south for the winter months.

This species adapts well to wildlife refuges and their requisite man-made, shallow impoundments. It breeds in both fresh and tidal wetlands, with scrapes situated on islands, dikes, or elevated platforms close to water, and less commonly on mats of floating vegetation. Pair bonding may occur on winter territories. Nest initiation is primarily late April to mid-May, with egg dates from early April to mid-August. Incubation is 18–27 days, and young can fly after about 28–32 days.

Black-necked Stilt is a partial to intermediate distance migrant, mostly through the interior, and they migrate primarily at night and early morning. Spring migration is from mid-February through mid-June, and fall migration takes place from July to December. Adults depart most breeding areas from early July to early September, often gathering in flocks at nearby staging areas in August. Southbound juveniles depart breeding locations from mid-July to early September and migrate mostly with adults.

After heavy rains or tidal flooding, Stilts may forage in roadside ditches and large puddles. Their primary prey is small aquatic invertebrates, and they will occasionally forage in water up to their breast. Prey is mostly detected visually and secured by pecking or plunging, although small invertebrates and fish are captured in deeper

Flying with Long-billed Dowitchers and Stilt Sandpipers, these wintering Black-necked Stilts at Llano Grande State Park in the Rio Grande Valley of Texas in November add a bit of flash to this flock of otherwise drab plumaged shorebirds.

water by scything their bills side to side in the water column, similar to avocets. Brine shrimp constitute an important food item in salt ponds. In freshwater aquatic and terrestrial environments, insects are important food sources, especially beetles, and small vertebrate prey (fish and tadpoles) are also sometimes eaten.

While Black-necked Stilt is unlikely to be confused with any other North American species, some authorities consider it conspecific with other stilts in the genus *Himantopus*, specifically Black-winged Stilt (*Himantopus himantopus*). No new information is available to adjust the estimate of 150,000–200,000 birds breeding in North America (Morrison et al. 2006), but Breeding Bird Survey trends are increasingly positive over the long and short term (Andres et al. 2012). Numbers for Hawaiian Black-necked Stilt's (*Himantopus mexicanus knudseni*) resident populations stand at 2,100 birds (U.S. Fish and Wildlife Service 2011 in Andres et al. 2012), and this number appears to reflect a large population increase.

▶ American Avocet
Recurvirostra americana

BIOMETRICS 17–18¾ inches long; wingspan: 29–32 inches; weight: 275–350 grams (9.8–12.5 ounces)
STRUCTURE Rounded body with attenuated rear; long legs and neck; small head; long, upturned bill (straighter and slightly longer in males)
STATUS Locally common through much of West and South; scarce in Northeast; population appears stable in long and short term

American Avocet takes the meaning of the word elegance to a higher level as it glides through shallow waters and swishes its rapier-fine, upturned bill side to side to catch aquatic invertebrates. A striking rust head and neck with black and white body/wings and bluish legs adorn its breeding plumage in summer, while winter finds it a stark black and white bird overall. Males have longer, straighter bills

◀ Is anyone having trouble understanding why this bird is called a "stilt"? This adult female Black-necked Stilt exudes grace and beauty as she stands at attention near High Island, Texas in April.

▽ American Avocet is the poster child for elegance. Even its name, derived from the Italian *avocet* or "graceful bird," underscores this assessment. Of the world's four avocet species, only American Avocet dons the rich, cinnamon blush in breeding plumage. Shown here are two males and one female (center), which has a noticeably stronger upturned bill. TEXAS, APRIL

Stilts and Avocets (Family Recurvirostridae)

▲ Striking in any plumage, these mostly nonbreeding birds have found the ideal water depth for roosting at the Salton Sea, CA in February. The right photo shows a juvenile bird in September in NYC that still has some remnants of pale buff-orange shading and patterns on its head, neck, and upper breast.

▼ Success! Even during copulation, these birds maintain poise and beauty! The male has to position his long legs on the back of the female to balance himself. Note the longer, straighter bill on the male in this special photo by Arthur Morris. APRIL

⚠ Small islands in freshwater ponds (*left*) are preferred nest sites for American Avocet, which they adorn with elaborate grass and stick nests. American Avocets nest in loose colonies, and their elaborate courtship displays conclude with a tender post-copulation display (*right*) that includes walking together gracefully with necks intertwined. SALT LAKE CITY, UTAH, MAY

with a slight upturn near the tip while females have slightly shorter and more strongly upturned bills.

American Avocets are stately birds of temporary wetlands in arid regions, and their preferred habitats for nesting include salt ponds, shallow alkaline lakes, and freshwater marshes and impoundments, where they often occur in loose colonies. They breed locally from s. British Columbia and sw. Ontario south to central Mexico, where they are resident. Distribution is closely tied to finding suitable water conditions in corners of the planet given to widespread seasonal variation in water levels. Courtship is an elegant affair, with sensitive posturing between birds that involves bowing and wing drooping, and males aggressively defend the pair bond from intruders, after which copulation occurs. Males have to position their long legs on the back of the female to accomplish the deed.

Avocets are typically seen in pairs or small groups at breeding sites. Birds may arrive on breeding grounds already paired and several weeks before nest construction, primarily mid-May. The nest is a scrape in soft, often elevated soil in a sparsely vegetated area close to the water's edge. Avocets also use natural or man-made islands, which may afford better protection from mammalian predators. Multiple pairs may nest in close proximity, particularly in times of drought or in places where water levels are managed. It is estimated that about 10,000 pairs of American Avocets nest in the vicinity of the Great Salt Lake every year (fide KK, 2012). Incubation is 23–30 days, with both adults sharing parental duties. Young can fly after about 27 days, and age of first breeding is 1–2 years.

In winter, these pied beauties gather on freshwater lakes and coastal flats from n. California south to sw. Guatemala, and from North Carolina south through Texas to Belize and Cuba. The large wintering population on the Bolivar Flats near Galveston, Texas has increased dramatically in the last 10 years, from 8,000 to about 12,000 birds, according to Winnie Burkett, a shorebird warden who has a house bordering the Flats. Here their massed ranks recall an M. C. Escher painting, with rows of black and white birds lining up in alternating synchronous fashion (see page 42, bottom photo).

Mostly a tactile feeder, avocets seine invertebrate prey from water through their swished bills, which are lined with small serrations. The birds wade like small herons, swim like ducks, dive like grebes, and rocket from the water like teal (Hall 1960). In waters along the Gulf of Mexico, avocets swim from sandbar to sandbar while plucking small fish and invertebrates from the shallow waters. Their flights over the Gulf's shallow waters are impressive, with lines of birds flying swiftly over water like arrows (see page 64, lower left photo).

Avocets are docile birds that rarely squabble with each other outside the breeding season, and their large flocks in winter move in almost hypnotic symmetry as they corral

△ Social at all seasons, American Avocets in winter and early spring may gather in the many thousands on tidal flats rich with dense concentrations of small invertebrate prey and fish. A large wintering flock at Bolivar Flats, Texas has grown from about 8,000 birds in 2008 to almost 12,000 today (Winnie Burkett, pers. comm.), as evidenced by this early April photo. Note the birds feeding in deeper Gulf waters in the background.

schools of small fish. They move in alternating rows while methodically feeding with their scapular (upper back) feathers spiked upwards in excitement (see page 42, bottom photo). When the word gets out that a feeding frenzy is nearby, thousands of avocets may descend upon the scene to join the fray.

Without appreciable fear of humans in relatively close proximity, it is a delight to observe these stunning birds en masse as they pursue their fish and invertebrate prey. Kevin has stood in shallow waters at Bolivar Flats, Texas in the middle of about 8,000 American Avocets for hours on end, totally in awe and at peace with these mesmerizing birds.

Once more widespread, with birds formerly breeding along the Atlantic coast, their future in present western strongholds is contingent on beneficial water use practices

▽ The elegance of avocets is accentuated in flight, as shown in this line of avocets flying swift and low like a pack of arrows over the Gulf of Mexico in Texas in April. The right photo shows a paired male and female flying over the Great Salt Lake in May.

▶ As thousands of American Avocets explode into flight, their plaintive piping calls collectively turn into a loud chorus of sounds that resemble a musical ensemble.

and water quality in the United States and wintering sites in North America and Mexico. According to North American Breeding Bird Surveys, their population remained stable between 1966 and 2015, with a population number of 450,000 in North America (Morrison et al. 2006). While Avocets historically adjust well to changing water conditions, permanent habitat loss resulting from altered land use practices may result in population declines.

▼ A school of small fish at Bolivar Flats, Texas in April has attracted a variety of excited birds ready for a good meal, including the obvious American Avocets. Snowy and Great Egrets and several large shorebirds (Western Willets and Greater Yellowlegs) join in the movable feast. Did you notice the dark morph Reddish Egret sneaking off at left?

Oystercatchers (Family Haematopodidae)

Numbering only 10 species globally, oystercatchers are large, mostly coastal shorebirds that live primarily in a narrow ecological zone of salt marshes, barrier islands, and rocky islands, where they feed primarily on shellfish. In stark contrast to sandpipers, oystercatchers are distinguished by lack of morphological divergence. Not only are the world's oystercatchers alike in size and structure, they are also similar in plumage. This fundamental similarity is underscored by all the oystercatchers occurring in the same genus, *Haematopus*, Greek for "blood foot," a reference to the pinkish legs of all adults.

All oystercatcher species are divided by birds that have entirely blackish bodies and those with a pied plumage, which exhibits a combination of bold black, brown, and white shading. North America boasts one of each: the American Black Oystercatcher of rocky Pacific coastlines, and the pied American Oystercatcher that occurs along the Atlantic and Gulf coasts and in extreme s. California and Baja California, where they inhabit coastal beachfronts and adjacent marshes.

Large, stocky, and somewhat neckless, all the world's oystercatchers brandish a straight, sturdy, laterally compressed, knifelike bill used to detach and decant mollusks, including assorted bivalves, oystercatchers' near-universal prey. The bright red eyes of Old World species and yellow eyes in New World oystercatchers give birds a maniacal or crazed expression, a sense underscored by their shrill, whistled yelps. Short, sturdy legs give birds a low center of gravity that help them navigate slippery rocks within the splash zone and also allow them to swiftly sprint away from danger or disturbance.

Foraging is somewhat tidally dependent, and on falling tides oystercatchers seek out exposed mussels, limpets, clams, and other bivalves procured by hammering, prying, or stabbing, depending on the prey. What oystercatchers truly seek are barely submerged mollusks whose shells are partially open to allow the mollusk to filter feed, enabling the oystercatchers an easier entry. The slightest vibration will prompt shellfish to snap shut, so feeding oystercatchers must navigate semi-exposed shellfish beds with slow deliberation. There is no need to hurry, as mollusks are hardly sprinters.

Along the Atlantic sandy coastlines, one of American Oystercatcher's favorite and staple food sources is the mole crab, a soft-shell decapod crustacean that burrows slightly below the surface. Mole crabs also occur on some Pacific sandy beaches, giving the Black Oystercatcher a valued meal that is relatively easy to secure and can be eaten whole because of its soft shell.

Oystercatchers are monogamous and highly territorial, vigorously defending their breeding territories from neighbors. All oystercatchers have a single brood per nesting cycle, with clutches averaging 2–3 eggs laid in a shallow scrape on the ground, which may or may not be lined (see page 28, lower photo). While physically precocial, chicks are fed by adults until well after fledging and may remain with parents for months while they master the highly specialized feeding techniques these jackhammers among shorebirds need to ply their trade.

Except for gulls, which access a mollusk's bounty by dropping them onto pavement, no other bird group is capable of feeding on large bivalves, a nutritious harvest for a highly specialized bird. Their specialized diet and

◀ Short, sturdy legs and a low center of gravity permit Black Oystercatchers to maneuver on slippery rocks, not to mention their large, partially webbed feet with large toenails.
ALASKA, MAY

▶ The long, sturdy bill of this American Oystercatcher is extracting a tasty meal from this Fiery Conch at Fort Myers Beach, Florida in January.

unique feeding methods do not advantage other birds except the occasional marauding gull that relieves unwary oystercatchers of their hard-earned fare.

Oystercatcher productivity is perilously low, with an average of fewer than one chick fledged per pair per year, and most mortality occurs before chicks fledge. While northern populations of American Oystercatcher are migratory, Black Oystercatchers are largely resident, with migration limited to postbreeding flocking. However, a large percentage of Alaskan Black Oystercatchers withdraw south in winter, with some moving as far as British Columbia. Postbreeding flocks of American Oystercatchers may number in the many hundreds before migrating south in the fall. In winter, the birds may gather in small flocks but do not join mixed shorebird flocks.

▶ Black Oystercatcher
Haematopus bachmani

BIOMETRICS 16¾–18¾ inches long; wingspan: 30¾–35 inches; weight: 500–700 grams (17.8–25 ounces)
STRUCTURE Bulky body with long, heavy bill; longish neck; thick legs and powerful feet
STATUS Locally common in northern areas of range; uncommon in south; strictly coastal

While many "black" birds are truly not, such is not the case with this sturdy denizen of North America's western rocky coasts. Were it not for the bright orange bill, red-rimmed yellow eye, and pink legs, Black Oystercatchers would be invisible against the backdrop of dark wave-washed rocks.

◀ After losing two hard-earned meals to this opportunistic Ring-billed Gull while trying to wash its conch snacks at the water's edge, this American Oystercatcher flew to a nearby island to enjoy its third catch without disturbance.
FLORIDA, JANUARY

▶ All oystercatchers have bright orange bills and yellow or orange eyes that impart to birds a crazed or maniacal expression. North America's Black Oystercatcher ranges coastally from the Aleutian Islands to Baja, Mexico and is almost never beyond the reach of the Pacific's tide.
ALASKA, MAY

▼ Oystercatchers have laterally compressed bills that allow them to pry open mollusks, similar to the operation of an oyster shucking knife. Once the bird inserts its bill into a gap in the mollusk's shell, the bird simply turns it to pry it open enough to cut the abductor muscle. This bird sports a look that leaves little to the imagination. If shellfish have nightmares, this is what they look like. ALASKA, MAY

Fresh-plumaged Black Oystercatchers from Alaska to Oregon are entirely black, while those along the extensive California coastline show increasing amounts of brown on the upperparts and belly, a result of occasional breeding with American Oystercatcher in the southern part of their range.

Mostly solitary or found in pairs, Black Oystercatcher's association with other wave zone specialists is mostly serendipitous. Advantaging Black Oystercatcher as it rambles over steep slippery rocks are gluelike pectinations on the bird's feet. While rocky coastlines are typically favored, birds also forage on adjacent sandy beaches. Oystercatchers' bills are laterally compressed, similar to oyster-shucking knives, which allows them easier access to slight openings in hard-shelled crustaceans.

▲ Talk about a room with a view, this Black Oystercatcher's nest scrape lies on the rim of its rocky perch and offers a magnificent view of Prince William Sound and its surrounding majestic mountains. ALASKA, MAY

Black Oystercatchers never seem hurried, as bivalves and their other food sources cannot run from foraging birds. Limpets, periwinkles, and mussels make up the bulk of their diet, and it takes young up to four months to learn how to open mussels and chip chitons and limpets from rocks. The implications of the name notwithstanding, oysters do not figure in this bird's diet, and most feeding occurs at low or changing tides.

The piping calls of Black Oystercatcher are similar to those of European and American Oystercatchers, with the male's call somewhat higher pitched. Vocalizations are typically given when pairs take flight, or when individuals approach or depart from scrapes. It is quite a moving sight to see a pair of Black Oystercatchers standing on a rocky shoreline in Alaska, dwarfed by the majestic mountains in the background while Black Bears amble along above them.

Ranging from the Aleutian Islands to central Baja, Mexico, these strictly coastal birds are mostly resident, and monogamous pairs tend to occupy the same territory year after year. Birds may vacate breeding areas in winter in

▼ At 1½ × 2 inches, eggs of Black Oystercatcher are large even by shorebird standards. The 2–4 eggs are usually seated in a shallow scrape on a pebble-strewn shoreline or on a rocky island, in this case shown in the previous photo. The broken shells are just window dressing for the nest (Alaska, May). A juvenile at right shows dark eyes and a dark-tipped bill in California in August.

▶ Noisy and animated display flights are typical in oystercatchers, illustrated by this pair of Black Oystercatchers flying over the Pacific Ocean in late July in California.

favor of nearby sheltered areas that offer high bivalve density, where they often gather with other Black Oystercatchers. In these locations, flocks may number in the hundreds. The southern limit of their range coincides loosely with the switch from rocky coast to sandy beach, although a number of birds occur in s. California due to man-made jetties providing the rocky habitat they prefer.

Black Oystercatcher is generally uncommon throughout its extensive range and not evenly distributed. Birds nesting on sand or gravel are subject to higher mammalian predation, and Glaucous-winged and Western Gulls are a threat to eggs and young. As with American Oystercatcher, their high, whistled calls are penetrating and carry great distances and are often accompanied by shallow, fluttering wingbeats.

Most egg laying takes place in May and early June, and one brood is typical, though birds will re-nest after failure. Both adults share parental duties, with incubation taking about 26–28 days. Young fledge after 38–40 days, but juveniles are still dependent on adults for food for at least a month after that. Family groups are maintained during the migration period and into early winter.

Despite more recent population studies at breeding sites and systematic surveys along the coastlines of Oregon, California, and Baja California, there was no need for revision of the previous population estimate of 10,000 birds (Morrison et al. 2006). Limited information available from Breeding Bird Surveys indicates a stable to increasing population trend (Butcher and Niven 2007 in Andres 2012).

▶ American Oystercatcher
Haematopus palliates

BIOMETRICS 16–17½ inches long; wingspan: 29–32½ inches; weight: 400–700 grams (14.2–28.5 ounces)
STRUCTURE Bulky, football-shaped body with long, heavy bill and longish neck; thick, relatively long legs
STATUS Fairly common at intertidal coastal habitats; northern populations migratory; uncommon to rare in s. California; population trends stable

This large, handsome coastal resident occurs almost exclusively in tidal wetlands and on sandy beaches but will move to rocky jetties and headlands in winter to access the bounty of mollusk beds, including small mussels and clams. Because of their specialized diet of mostly shellfish and other saltwater crustaceans, oystercatchers live only in a narrow ecological zone between salt marshes and barrier beaches. They are among the most visible and vocal members of sandy coastal communities, and their almost comical appearance is unforgettable to all who view them.

Two races breed in North America: the nominate race extending from Massachusetts south to Georgia, and a race that extends from Florida's panhandle west and south to extreme northern Mexico. In Audubon's time the birds ranged north to Labrador. Birds are resident from Maryland southward, although some individuals remain in s. New Jersey and Delaware for the winter. A Pacific race breeds coastally from Baja California and the Sea of Cortez south to Panama. Local resident populations also occur on

American Oystercatchers are large, chunky birds with long, sturdy bills and thick legs that lack hind toes. Found mostly in sandy coastal areas, they do not choose to associate with other shorebird species. Note the black designer toenails on this bird. NJ, SEPTEMBER

American Oystercatcher courtship and copulation is energetic and vocal. Pairs mate for life, and the 1–4 cryptically patterned eggs are deposited in a shallow scrape on open sand, cobble beach, or tidal marsh (hopefully above the high tide line). The copulating birds are on the Galápagos Islands in October, where this species is resident, while the nesting bird is at Stone Harbor, NJ in May.

the Galápagos Islands, the Pacific coastline of South America, and Atlantic coastlines from Brazil to Argentina.

In winter, northern Atlantic birds occur in small to large flocks as far north as New Jersey, and there is some evidence they may be trying to reestablish their historic range, with recent breeding records from Maine and Nova Scotia. The distribution and abundance of this species appears inexorably linked to the presence of shellfish beds accessible at low tide. The name "oystercatcher" evidently harks back to naturalist Mark Catesby, who observed American Oystercatchers foraging over oyster beds near Charleston, South Carolina in the early 1700s.

This bivalve specialist does indeed include oysters in its diet, along with mussels, clams, crabs, sea urchins, starfish, and marine worms. Another favorite food staple along Atlantic sandy beaches is the mole crab, a softshell, saltwater decapod crustacean that burrows slightly below the surface in the intertidal zone, and whose presence is revealed to oystercatchers when the outgoing waves cause air bubbles to drift to the surface. The long bill of American Oystercatcher probes near the bubbles and more often than not comes up with a substantial, protein-rich food prize. In addition to stabbing or prying into partially opened bivalves, American Oystercatcher sometimes hammers

▶ Owing to their highly specialized diet and feeding techniques, chicks are tended by adults for 3 or more months. The teenager on the left hears the dinner bell and is receiving a morsel of mole crab, one of their favorite surf zone foods.
NJ, LATE JULY

its way into shellfish, cutting the shell-closing abductor muscle once entry is gained.

Monogamous pairs typically remain faithful to breeding territories for multiple years. Incubating or brooding adults take to the air grudgingly, preferring to run swiftly away from the nest to distract potential predators, a response quite unlike the histrionic displays of Eastern Willet, which may be nesting nearby, and which start calling when observers are a hundred yards away (see page 247, upper photo).

Oystercatchers may place their nests within tern and skimmer colonies for the air cover they afford, but it may be the prime coastal real estate above the reach of the high tide that is the true attraction. This species readily nests on dredge spoil islands and even partially submerged shipwrecks, even though these man-made foundations are equally attractive to Herring and Great Black-backed Gulls, which may prey on oystercatcher eggs and young.

When approached by potential danger, flightless oystercatcher chicks play dead and hope the potential predator is fooled by their inaction. Their natal plumage blends perfectly with the sand, and this is another reason for the chick's "playing dead" behavior (see page 31, lower photo). From the sky, they are virtually invisible to aerial predators such as gulls if they don't move. Young oystercatchers are able to stand and run within hours of hatching, but

▼ The left photo shows American Oystercatchers posturing near the surf line to warn other oystercatchers to exit their territory. During winter months, mussel colonies on rocky jetties supplement the diet of American Oystercatchers. NJ, JULY (*LEFT*); DECEMBER (*RIGHT*)

▲ From spring to fall, American Oystercatchers engage in spirited and noisy display flights with neighboring pairs that may include up to 10 birds. These flights not only are part of the breeding territorial behavior but also occur out of season just for the sake of posturing to their neighbors when multiple birds are concentrated together. NJ, JUNE (*LEFT*); APRIL (*RIGHT*)

▼ These very young chicks are about to receive a tasty mussel dinner. Because of the complex feeding protocol involved in prying into mollusks, chicks have to be fed by adults for several months until their bills fully form. NJ, MAY chicks are dependent on adults for food for up to 4 months. Acquiring shellfish is easy; decanting them takes training.

One of the most unique and compelling behaviors of American Oystercatcher is its propensity to fly in synchronous fashion with up to eight neighboring oystercatchers in a pattern that would make the US Air Force Blue Angels envious. This flight display is accompanied by noisy screaming and fast, steady motion that can last up to several minutes. While these flights are often referred to as courtship displays, and this may be accurate, they are also performed before and after nesting is completed and are simply displays of social posturing.

These posturing flights are often instigated by a single pair of birds, and they spread to adjacent areas where other American Oystercatchers are magnetically attracted to the growing pack of screaming adults. Even brooding and incubating adults will abandon their eggs or chicks to engage in these social displays, after which they return to their nests or brood sites like nothing ever occurred.

While most shorebirds never leave their young unattended and defend them with aggressive behavior, it is interesting that a posturing display with neighbors carries more weight to American Oystercatcher than staying with its eggs or chicks, at least for several exciting minutes.

Only the exquisite camouflage of downy young prevents more predation from aerial predators, as they literally blend into the sandy substrate that is their home.

Nest initiation is primarily early April in the south and mid-May in the north. Incubation of the typical 2 or 3 eggs is 24–27 days, and young are able to fly in about 34–37 days. Both adults share parental duties, and one brood is typical, but they may re-nest after failure. This species is largely resident, but birds that breed from Maryland northward are mostly migratory. Short-distance migration takes place along the coast and over littoral waters, primarily during the day.

American Oystercatcher has one of the swiftest and most powerful flights of any shorebird, and it can achieve rapid speeds in a very short time. When one of these large shorebirds races by you on a coastal beach below eye level, you can hear the rush of its wings and almost feel the wind from their flight.

Habitat loss and climate change are distinct perils to this species as sea level rise threatens to wash out more nests every year. No new information is available to refute the population estimate of 11,000 (Morrison et al. 2006), and Christmas Bird Counts indicate a stable population trend (Butcher and Niven 2007 in Andres et al. 2012).

▼ Beginning in July for failed breeders and later for successful adults and juveniles, American Oystercatchers gather in postbreeding aggregations at coastal areas. These birds are staging in the shadow of New York City in September.

Plovers (Family Charadriidae)

Plovers are stocky, round-headed, somewhat neckless birds with legs designed more for walking and running than for wading. Bills are mostly short, blunt, and bullet-shaped, ideally suited for plucking prey from the surface. All plovers are remarkably cryptic, but moving across open terrain, they are potential prey to a predator's hunting eyes. Immobile, they melt into the substrate: sand, prairie, tundra, or gravel bar.

Mountain Plover, a sand-colored prairie breeder, literally disappears into the landscape simply by turning 180 degrees and concealing its whitish underparts, a maneuver that earned this North American endemic the nickname "prairie ghost."

There are few sights and sounds in the bird kingdom more moving than plovers navigating their breeding territories on the tundra. The striking plumages of the large North American plovers range from golden hues to bold black-and-white contrasts, and their whistled calls are both plaintive and evocative.

Plovers are small- to medium-sized shorebirds that forage in a wide variety of habitats, including tundra, open wetlands, tidal flats, sandy beaches, moist to wet meadows, and short (often arid) grasslands. While the largest North American plover is only slightly less than 2 times longer than the smallest, they appear larger because of their substantial bulky bodies and longer legs.

These dedicated sight hunters with unusually large eyes use a start, stop, and tilt foraging strategy and an upright stance between steps to assist with prey detection. In contrast to plovers, sandpipers are mostly tactile

▼ Plovers, like this Black-bellied and bracketing Semipalmated Plovers, are distinguished by their short, blunt-tipped bills, chunky bodies, and short legs. Black-bellied, our largest plover, measures twice the length of Semipalmated, but it appears much bigger because of its bulky body, large head, and long neck. NJ, MAY

Wilson's Plovers often stand near driftwood or beach vegetation for the protective cover they afford in the wide-open habitats frequented year-round by this species. This is one of Kevin's favorite shorebird photos because of its artistic nature, not to mention the gorgeous shell beach at St. Augustine, Florida, April.

Semipalmated Plovers are very aggressive defenders of small feeding spaces during migration. They typically face off and posture with heads down and tails fanned, and sometimes one will leap on top of the other to gain the upper hand, after which they walk away gingerly while casting nervous glances at each other. NJ, MAY

probers, feeding on the run and jabbing as they move. Plovers walk, stop, and pick. The larger plovers cover more ground and detect prey farther away than smaller ones, and they also consume larger food items.

As dedicated hunters, plovers are somewhat less social than sandpipers, spacing themselves farther apart when foraging. Some birds vigorously defend feeding areas in migration and on the wintering grounds by confronting transgressors with threat displays that involve charging with head lowered and tail fanned. Following the initial challenge, confrontational birds may walk side by side along the respective undefined boundary line until peace is restored.

Because they feed day and night, plovers' large eyes are rod-cell rich, facilitating the detection of motion. Some species also employ a foot pattering maneuver where one foot taps the moist substrate, either to force prey (mostly terrestrial and marine worms) to move, thus disclosing their location, or to coax prey to the surface. Foot trembling often results in higher capture rates (see page 105).

While plovers are a large family group, only 11 of the planet's 67 plovers breed in North America. Four of these (Black-bellied, American and Pacific Golden-Plovers, and Eurasian Dotterel [rare]) are tundra breeders and medium- to long-distance migrants. All have plaintive, whistled flight calls that invoke in human listeners a longing for travel to lands beyond the horizon. Pacific Golden-Plover is one of the planet's long-distance champions, flying nonstop from Alaska to Hawaii, nearly 3,000 miles.

The closely related American Golden-Plover is believed to fly nonstop from James Bay, Canada, to Argentina, a staggering distance of 6,500 miles. Migrating alone or in small to large flocks, the group composition of flocking plovers never approaches the density of sandpipers. Eurasian Dotterel is an Old World tundra breeder that ranges from Norway to Siberia and at widely scattered high-mountain sites from Italy to Mongolia, but it extends its breeding range to the Seward Peninsula of Alaska north to Point Barrow (Nuvuk). Dotterel winter in Spain, North Africa, and the Middle East.

Piping, Snowy, and Wilson's Plovers breed well south of the tundra in temperate areas and winter coastally in the United States, Mexico, the Bahamas, and the Caribbean. Semipalmated, the fourth common small plover in North America, has the widest distribution of all North American plovers, breeding in the northern tier from Alaska to Newfoundland, and occurring in every one of the lower 48 United States in migration before wintering on all s. US coastlines south to coastal areas of s. South America. A fifth small plover, Common Ringed Plover, is

a mostly Eurasian species that breeds in small numbers (about 2,000 birds) in remote barren regions of n. Canada, but it is rarely seen in North America away from breeding areas.

Two species, Killdeer and Mountain Plover, breed and winter in interior as well as near coastal portions of North America. Killdeer is widespread and breeds from coast to coast, while Mountain Plover is localized on the High Plains (breeding) and Western prairies, wintering mostly in California, Arizona, Texas, and northern Mexico.

Plover chicks are precocial, with adults and young leaving the vicinity of the nest within hours of hatching. While chicks can feed themselves, parental care is essential for 2–3 weeks to protect juveniles from predators and to shade young in warmer climates and warm them in colder regions. In early stages of chick development, this results in comical scenes where brooding parent birds appear to have five sets of legs as the chicks crowd under their wings and bellies (see page x). Plover chicks develop more

▲ This cute 2-day-old Piping Plover chick is mostly legs with a puffball on top, but it lives a tenuous existence on the open beach habitat where it hatched. Adults have to exert special care for these young chicks to protect them from predators such as gulls, terns, foxes, and large ghost crabs that want an easy meal, and from the hot sun and bad weather. NJ, MAY

slowly than sandpiper chicks, and their slower metabolism may benefit the birds in times of food shortage.

Because of the open habitat favored by plovers, the survival rate of chicks is lower on average than that of sandpipers. If young can avoid starvation and aerial and terrestrial predators, they fledge after 21–32 days on average for smaller plovers, and 21–28 days for American and Pacific Golden-Plovers. The young of Black-bellied Plover take longer to achieve flight, averaging 35–43 days. The smaller plovers ("ringed plovers") may live for 10 or more years (though most do not), with Snowy Plover averaging about 3 years and Piping Plover almost 5 years, although the oldest Piping Plover on record lived 14 years in the wild, and the oldest Snowy was 15 years and 9 months, suggesting that life span data for these species are inaccurate and need more work. The larger species, such as Black-bellied Plover, may live for 20 years or more in the wild.

As to the origin of the name "plover," there seems no accepted source, although *pluvialis* is Latin for "rainy," which might imply that those humans who hunted plovers noted a link between numbers of grounded plover migrants and rainy weather.

Plovers appear to have evolved in arid, semi-desert conditions, and most species occur in the Southern Hemisphere. Their adaptation to cold tundra environments as migratory breeders is a testimony to their adaptability and hardiness, and their long-distance migrations are a credit to the evolutionary process that always seems to find a way.

▶ Black-bellied Plover
Pluvialis squatarola

BIOMETRICS 10¾–12 inches long; wingspan: 29–32 inches; weight: 105–320 grams (6.5–20 ounces)
STRUCTURE Chunky and compact shape with thick neck, bulky chest, large head, and heavy bill. Primaries extend only slightly past tail tip.
STATUS Common in coastal areas outside breeding season; less common inland during migration. Worldwide distribution; known as Grey Plover outside the United States.

This large, burly plover is a fixture on winter beaches and mudflats on all coasts, as well as on moist inland cropped grasslands and fields. Its whistled contact call (*pee-oo-EEE*) is as much a part of the winter soundscape as the lap of waves and the keening cry of Herring Gull. Black-bellied is also one of the most widespread of all shorebirds, occurring on six continents around the world, including almost all coastlines of North, Central, and South America, but it is called Grey Plover worldwide except in North America.

Whether escorting a Glaucous Gull out of its breeding territory or racing across a subtropical tidal flat, this silver backed, onyx-breasted "King Plover" rarely fails to impress with its poise and presence. It is the largest plover by far in North America, with the biggest Black-bellied weighing about 65% more than the smallest American Golden-Plover.

Less social than the golden-plovers, Black-bellied was less impacted by market gunning, and the wariness of the

▶ This large, robust, silver-backed, onyx-breasted Black-bellied Plover breeds in a variety of Arctic tundra habitats in the Old World and New. The mostly white "snow cap" on this male is usually seen only on the breeding grounds, which in this case is Alaska's North Slope. ALASKA, JUNE

birds was both cursed and celebrated by gunners, who regarded the "bullhead" a prize. Other shorebirds also appreciate the attention to potential danger by Black-bellied, and its sudden taking to flight with emphatic whistled call notes brings up the heads of all feeding shorebirds, if not spurring them to instant flight.

Black-bellied uses this enhanced mass to nest and winter farther north than most shorebird species, with birds breeding from w. Alaska to the inhospitable islands of the Canadian Archipelago north to Devon Island (75°40'N). In winter, they range coastally from British Columbia (rarely s. Alaska) and Massachusetts (rarely Newfoundland) south to s. South America, the West Indies, and the Bahamas. Two subspecies breed in North America: *P. s squatarola* in Alaska and *P. s. cynosurae* in the Canadian Arctic.

Spring migration begins in early April, and migrants are most common in coastal areas. Peak numbers pass through the Gulf and Pacific coasts in late April, and the s. Alaskan and mid-Atlantic coasts in May. Some 1-year-old birds remain on the wintering grounds through the first summer, while others migrate partway or all the way north to the breeding grounds; thus small numbers of birds are regularly seen through much of the species' range in midsummer.

On the high Arctic tundra breeding grounds, this sturdy plover is a fierce protector of its large nest area, and it strikes fear into aerial predators as large as gulls and jaegers. Woe to any flying predator if it hears the whistled alarm call of Black-bellied Plover as it flies swift and low across the tundra before rising to spear the intruder in the belly with its strong bill (see illustration by Sophie Webb on page 27).

Kevin regularly witnessed other shorebirds, particularly Buff-breasted Sandpiper, choosing to nest near Black-bellied in the Alaskan Arctic for protection against predators, where only the female Buff-breasted is present during the nesting and brooding periods. This aggressive plover has no equal when it comes to nest protection and determined physical defense.

First eggs are laid from early to mid-June in northern regions, but somewhat earlier in southwestern Alaska. Owing to the vigilance and determined defense of adults, predation on nests is low, although Arctic Foxes pose a persistent challenge, exacerbated in years when lemming numbers falter. The nest of Black-bellied is a scrape in sparsely vegetated habitats, including moist grassy tundra or raised, dry polygonal mounds, with males often lining their nests with lichen and other leaves, such as Buckwheat, to impress the female.

Plovers like an unencumbered view of the nearby landscape and are often partially exposed as they incubate. During this time, they regularly lower their heads on the nest to limit their exposure. When nests are close to hatching, birds often engage in animated nest distraction displays to lure potential predators away, including the "rodent run" behavior, when birds run along the ground while flapping their wings as if injured.

Incubation is about 27 days, with peak hatching late June to mid-July. While precocial, young tend to remain in or

◀ In a shallow pool in NJ in May, a stunning male Black-bellied Plover pauses between steps to search for food, all in the fashion of American Robin's foraging style. NJ, MAY

▼ This lone male Black-bellied Plover was the first shorebird Kevin spotted on the tundra after heavy snow melted on Alaska's North Slope in 1993 in early June. The "King Plover" is surveying his prospective breeding territory. The right photo shows a female Black-bellied settling onto her tundra nest in Alaska in June.

▶ Distraction displays like this "injured bird" posture are intended to draw encroaching predators (or shorebird technicians) away from eggs and nest. Black-bellied usually does this type of distraction display when the eggs are close to hatching. ALASKA, JUNE

close to the nest longer than most shorebird species and are brooded by adults for 2–3 days, a testimony to the harsh high Arctic weather. When chicks disperse, both adults brood separated young simultaneously, but the female deserts the family group after about 12 days. Otherwise, young fend for themselves, relying on parents to guide them to feeding areas, keep them warm, and protect them from predators.

Juveniles fledge or attain flight after 35–43 days, a very long time compared with sandpipers and other plovers. The prolonged nest and brood period results in fairly late fledging by young, but with snow covering the ground in some of their nest territories by mid-August (depending on latitude), they have to be careful not to nest too late in the season.

In migration and in winter, coastal birds forage in the intertidal zone but above the immediate tide line (i.e., they are not wave chasers). In the Midwest, migrating birds are drawn to plowed fields and the grubs and earthworms that are exposed. Southbound migration of adults begins by mid to late July and continues into December. Birds migrate on a broad front but are more abundant in coastal areas. In the Bay of Fundy, it is the only shorebird as common in spring as in fall, suggesting that individual birds use similar migratory routes, not the elliptical route favored by American Golden-Plover.

Winter range includes coastal areas from s. Canada and the United States to s. South America. Birds defend winter territories rich in marine worms, small bivalves, and small crabs from other plovers, but they may feed in

▽ A typical coastal scene as Black-bellied Plovers in assorted plumages and other beach foraging shorebirds sit out the high tide on an elevated sand spit in NJ during late August migration. Other shorebirds include Western Willets, Ruddy Turnstones, and American Oystercatchers.

While most golden-plovers elect to winter in warm climates, some hardy Black-bellied Plovers winter on the beaches of North America as far north as Alaska and northern Quebec. Birds typically roost in small to large groups, except after storms, when hundreds of birds may gather together. Also present are Dunlin, which occupy the same beachfront habitat and often roost with Black-bellieds.
NJ, LATE NOVEMBER

A striking male Black-bellied Plover is shown in flight in May with black wingpit feathers spiked outward. The right photo shows a bird molting into breeding plumage in March in Santa Barbara, California, with its shadow adding an artistic slant to this stark scene.

the company of Red Knot and Dunlin and typically ignore the presence of nearby sandpipers. Not really a joiner, Black-bellied Plover forages alone or in loosely spaced flocks of 20 or more birds. Roosting flocks typically number fewer than 20 individuals, but they may approach several hundred birds at times during storms or at favorite stopover locations. Red Knot and Dunlin especially rely on Black-bellied to warn of approaching danger in migration and winter.

Substantial declines of Black-bellied occurred on the winter grounds of n. South America between the 1980s and 2010s, as did migratory counts along the Atlantic Coast from the 1970s to 1990s, but stable or increasing counts have occurred since then. The overall estimate of Alaska's *P. s. squatarola* subspecies is 262,733, but this number is probably conservative based on new information from Arctic PRISM surveys. Estimates of Canada's *P. s. cynosurae* subspecies are 100,000 birds, bringing the total North American breeding population to 362,733, and population trends appear to be stable (Andres et al. 2012). World population numbers stand at 692,000.

GOLDEN-PLOVERS

Until 1993, the two North American Golden-Plovers were considered a single species called Lesser Golden-Plover. However, the breeding areas of what we now call American Golden-Plover and Pacific Golden-Plover are mostly segregated, and the wintering areas are completely separate and reached by mostly different migratory routes that link the birds to entirely different continents.

But the breeding grounds of the two species do overlap in western Alaska and western Siberia, where the birds intermittently interbreed, as evidenced by a bird in Colorado in the early 2000s that mostly resembled a Pacific Golden in plumage, but the DNA of a retrieved tail feather showed the mother to be American Golden. Another bird from Cape May, NJ in mid-October 2004 and documented in the hybrid section of *The Shorebird Guide* (O'Brien et al. 2006) showed plumage and structural features of both species and gave call notes of both species (fide Michael O'Brien). While hybrids of these two species are not common, it is difficult for most birders to spot differences in plumage or physical features in the field that suggest a hybrid.

◢ This hybrid American Golden/Pacific Golden-Plover juvenile spent a week on the beach in Cape May, NJ in October 2004, where it gave call notes of both species and showed physical and plumage features of both species. The tug of war with this marine worm was easily won by the bird.

◢ Young juveniles of both American Golden and Pacific Golden-Plovers are shown in late August in Nome, Alaska, with American Golden the darker brown ones with unpatterned underparts. While these species breed close to each other on the Seward Peninsula of Alaska, interbreeding rarely occurs.

Where the two species occur together, they use different breeding vocalizations and may select different nesting habitats, but a photo by Brian Sullivan shows juveniles of the two species together on the Seward Peninsula in Alaska. Where breeding ranges do not overlap, the species are less discriminating.

Both species rank among the planet's long-distance migration champions, and both are capable of high-speed flight, with speeds up to and perhaps exceeding 100 mph. Other estimates are more modest, and it is unknown how long birds can maintain these excessive speeds.

▶ American Golden-Plover
Pluvialis dominica

BIOMETRICS 9½–11¼ inches long; wingspan: 23½–28¾ inches; weight: 100–200 grams (3.5–7.1 ounces)
STRUCTURE Slender but athletic with rounded belly, long wings, and attenuated rear body; smaller head, smaller pigeon-like bill, longer legs, and slimmer neck and chest compared with Black-bellied Plover
STATUS Locally common in Great Plains and Texas coast in spring; less common through Central flyway; uncommon in Northeast in fall; much scarcer on coastal areas than Black-bellied Plover

- -

This elegant and delicate large plover elicits admiration among birders not only because of its mystique as a super long-distance global traveler but also for the bird's stunning breeding plumage, which rivals any shorebird for sheer beauty.

Breeding exclusively in tundra regions of North America, this highly social, gold-spangled, long-distance migrant was nearly hunted to extinction along its Atlantic (fall) and central Interior (spring) migration routes to and from wintering areas in the Rio de la Plata grasslands of south-central South America. Flying in flocks of a hundred or more birds, American Golden was easily drawn in by decoys and a favorite target of market gunners.

Its equally common migratory companion, the Eskimo Curlew, was sadly hunted to extinction. Today, American Golden-Plover numbers have recovered somewhat, but they have not returned to pre-market gunning numbers, and tragically the birds continue to be hunted for sport on Barbados and a few other West Indian islands.

It is estimated that 8,000 birds are killed annually in Barbados, and while only a fraction of the 48,000 birds estimated by John James Audubon slaughtered in a single day in New Orleans (Audubon 1827), the gunning in Barbados is still a sad commentary on our resolve to protect migratory shorebirds. Nevertheless, with a North American population estimated at a healthy 500,000 individuals, the birds seem able to shrug off the current rate of attrition.

Spring migration takes place between February and May, with arrivals along the Gulf Coast beginning in late February and peaking from mid-March to late April. Much of the spring migration takes place in the central part of the United States, with smaller numbers passing along the Atlantic and central Pacific coasts.

American Golden-Plover breeds in subarctic and Arctic regions from western Alaska to Baffin Island and Quebec province. Birds arrive on far northern breeding grounds

▶ The star-spangled American Golden-Plover must rank among our most stunning shorebirds. Adult males like this show a high degree of fidelity to nest sites, often returning to the same patch of tundra year after year.
CHURCHILL, MANITOBA, JUNE

from late May to early June, with much of the tundra still blanketed by snow. Unpaired birds seek out wind-blown hilltops free of snow where pair bonds are established. Pairs are seasonally monogamous. Both sexes incubate the clutch of 3–4 eggs and defend territories. Males frequently return to the same territory each year, but females are more fluid in their choice of territories and mates.

While working as a shorebird biologist in Alaska, Kevin found a male American Golden-Plover sitting on a nest in mid-June and went to mark the nest with a colored tongue depressor. When the bird flushed from the nest cup, he saw that there were already two worn tongue-depressors near the nest, which caused him to think that a fellow biologist had forgotten to log the nest into the data sheet.

After retrieving the depressors, he found that they represented the last two summers of nesting, and that the male plover had returned to the same square meter of ground among 178,000 miles of similar habitat on Alaska's

▲ Female American Golden-Plovers are less brightly colored than males, and they may reject several nest sites built by the male before choosing one. This well-nourished bird is shown on beautiful dry tundra in Point Barrow, Alaska in June. Stringy, white *Thamnolia* lichen on the ground is often used to spruce up the nest cup.

▼ American Golden-Plovers place their nests in locations where they can survey the surroundings for potential danger, and this beautiful male is nestled into soft tundra in Churchill, Manitoba in June. A male at right is performing a broken-wing distraction display to lure assumed predators like Kevin the working biologist away from its nest site in Alaska, June.

The superb camouflage of American Golden-Plover also extends to eggs and chicks. Four very recently hatched stunning golden chicks are still helpless, but they show the success of synchronous hatching, in which all hatch within a short time of each other, even though it took the female 6–7 days to lay all 4 eggs. Can anyone say Charles Darwin?
ALASKA, JUNE

North Slope after flying about 9,000 miles from South America. This is a true testimony to the pinpoint accuracy of the GPS system built into long-distance shorebirds.

The male may have to build another nest if the female does not approve of the first effort, or if she feels the nest is not properly camouflaged. When Kevin found a female American Golden-Plover about to lay the first egg in a nest, she abandoned the nest and prompted the male to build her a new one, which Kevin found the next day. These large Arctic breeding plovers also fancy putting their nests near driftwood in open coastal tundra within sight of the Beaufort Sea, as it gives them added cover.

Eggs and young are preyed upon by an array of avian and mammalian predators, with nest failure particularly acute when lemming numbers are low. Incubation is 26–27 days, and both adults share parental duties, though males tend the young longer than females. When young are capable of flight (21–22 days), one or both adults depart and migrate, or one parent (usually the male) may accompany young to form foraging flocks with neighboring birds.

Adults depart the breeding grounds from late June (failed breeders) to early August, and most depart e. Canada and head south on a direct flight over the Atlantic to wintering grounds in South America. Fall migration occurs through

During migration, earthworms play a major part of American Golden-Plover's dietary protein. This molting adult on Jones Beach, NY in September is winning the age-old contest between plover and worm with a hard-earned but substantial meal.

Fall migration for American Golden-Plover extends from late June to October, but prior to the International Migratory Bird Treaty of 1918, migrants ran a gauntlet of market gunners from Newfoundland to Barbados. This juvenile (*left*) from New Jersey in September is lucky to have hatched in more enlightened times, though these birds are still relentlessly hunted for "sport" in several West Indian islands, especially Barbados. A muted-plumage nonbreeding bird is shown at right from Texas in March.

late November, although most birds depart North America by mid-September. Juveniles typically begin their trans-Atlantic migration of 6,000–8,000 miles without accompanying adults by mid-August, with stragglers remaining into September.

In all seasons, the main food taken is insects and their larvae, spiders, small mollusks, and crustaceans. They also eat berries, seeds, and leaves that constitute important food items in spring and fall. Crickets and earthworms are regarded as particularly important autumn food items in coastal areas, as are grasshoppers along the bird's return route in the Central Plains of North America in spring. This elliptical migration pattern is used by other shorebird species, including Buff-breasted and Pectoral Sandpipers, and formerly Eskimo Curlew.

New information from Arctic PRISM surveys indicates a much larger population than previously thought. Bart and Smith (2012, in Andres et al. 2012) estimated the population in Alaska at 282,249, and in Canada at 208,570, and the Wader Study Group Bulletin titled "Population Estimates of North American Shorebirds, 2012" (Andres et al. 2012) suggests a number of 500,000, which is more than double the number previously proposed by Morrison and colleagues (2006). Even this number may be conservative, as greater than 40% of the breeding range in alpine regions of subarctic and boreal biomes occurs in areas not covered by Arctic PRISM surveys. Current population trends are unknown.

▷ Pacific Golden-Plover
Pluvialis fulva

BIOMETRICS 9¼–10½ inches long; wingspan: 21½–24½ inches; weight: 100–200 grams (3.5–7.1 ounces)
STRUCTURE Similar to Black-bellied Plover with large, blocky head; more front-heavy and upright stance than American Golden with a rounder body, larger head, heavier chest, and often larger bill and longer legs. Also shorter wings than American Golden, typically with only three primaries extending past tertials.
STATUS Uncommon during migration and winter along Pacific Coast; rare elsewhere

Dazzling in breeding plumage with jet-black breast and face and bold, golden-spangled upperparts, Pacific Golden can easily claim the title of most attractive breeding plover species in North America. These graceful, long-winged shorebirds perform animated, butterfly-like courtship flights on their tundra breeding grounds, similar to those of American Golden-Plover, and they drive most other shorebirds from their breeding territories, with the exception of Dunlin. Because of this strange relationship, the Icelandic name for Dunlin translates as "plover slave," while in 19th-century England, Dunlin was referred to as "plover page."

Discovered in Tahiti in 1773 on Captain James Cook's second voyage, this medium -sized tundra breeder ranges

◀ It's hard to decide which is more stunning: the breeding male Pacific Golden-Plover or the breathtaking colorful tundra in Point Barrow, Alaska in June. A tie would be the wise choice. In 1993, the American Ornithological Union split Lesser Golden-Plover into two species: American Golden-Plover and Pacific Golden-Plover. Kudos to Jamie Cunningham for this striking photo.

from the Arctic regions of Russia to western Alaska. It undertakes one of the world's longest migrations, flying from Alaska to the islands of the South Pacific, where it forages on mud, sand, pastures, golf courses, and air-fields.

Allowing for a brief stopover in Hawaii, where a fair number of Pacific Goldens overwinter, many birds travel on to Australia and New Zealand, an additional minimum distance of 3,000 miles over open water. In total, they migrate more than 8,000 miles during this long journey.

Birds from Russia migrate primarily to Asia, Africa, and Australia, with Siberian birds migrating south along the Asian coastline. Meanwhile, a small number of Alaskan breeding birds move south to winter in coastal areas of California and Baja Mexico.

While the breeding range of Pacific Golden-Plover is large, the wintering range is extensive, covering nearly half the earth's circumference. It extends from California to Hawaii, Asia, northeastern Africa, Australia, and New Zealand. It is unknown how much of this range is

▶ While most of North America's Pacific Golden-Plovers winter in Australasia, a few like the nonbreeding bird pictured spend the winter along the Pacific coast from California to Mexico, in this case near the Tijuana River mouth. Long legs and bill, large blocky head, and chest-heavy body structure help to separate this species from American Golden-Plover in winter. FEBRUARY, CA

occupied by North American birds, but it seems likely that they use winter ranges similar to those of Bar-tailed Godwit and Bristle-thighed Curlew.

Spring departure from wintering areas begins as early as February in southern parts of the range, and in early to mid-April in California. Breeding plumage is attained considerably earlier than American Golden, which helps in spotting vagrant Pacific Goldens in flocks of American Goldens in Texas and the Great Plains in April. Arrival on the breeding grounds takes place from April at southernmost sites to mid-June at high latitudes.

Pacific Golden breeds primarily in tundra with dense vegetation and a few rocks. Egg laying occurs from late May to mid-June, with Pacific Golden preferring lower, more shrubby tundra than American Golden. Incubation is 25 days, with males sitting on the nest during the day and females at night, similar to American Golden-Plover, but this scenario is hard to explain since the sun never sets during incubation season in the high Arctic. Emergence of all four chicks occurs within one 24-hour period (synchronous hatching) as a result of the adults not incubating the eggs full time until the entire clutch is laid, similar to all North American shorebirds. Young fledge in 26–28 days.

Both parents attend chicks, although males usually attend chicks longer than females. Young are able to run

▲ Two Pacific Golden-Plovers are transitioning into breeding plumage, with the left bird from Santa Barbara, California in March, and the right bird from North Carolina in April, where it appeared as a vagrant in this messy plumage in 2021 and 2022. Thanks to Audrey Whitlock for the second photo.

◄ While not as striking in plumage as the male, this female Pacific Golden-Plover carries her subtle beauty with style in Point Barrow, Alaska in June.

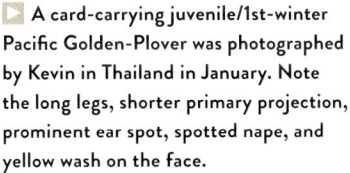 These young juvenile Pacific Golden-Plovers from late August in Nome, Alaska show the classic bold, golden-spangled upperparts, yellowish face, and buffy underparts with coarse streaks on the upper breast. Thanks to Brian Sullivan for this great photo.

 A card-carrying juvenile/1st-winter Pacific Golden-Plover was photographed by Kevin in Thailand in January. Note the long legs, shorter primary projection, prominent ear spot, spotted nape, and yellow wash on the face.

soon after hatching, with the first hatched chicks tending to feed around the nest while one adult continues to incubate late-hatching eggs. Both adults then lead young to wetter tundra where foraging success is higher and, more importantly, there is additional vegetative cover. Adults depart breeding areas from late June (failed nesters) to mid-July, and sometimes to early August, at which time flocks of fledged juveniles form on the tundra, especially in coastal areas. Juveniles migrate south in mid to late August, with stragglers remaining to early October.

Pacific Goldens migrate day and night in large flocks. Some birds from the eastern part of their range overwinter in Hawaii, but most are presumed to fly nonstop to their southern Pacific and Indian Ocean winter range, or perhaps with brief refueling stops at the Marshall or Phoenix islands.

While population trends appear to be downward, the overall population of 170,000–220,000 birds is sufficiently large to allay immediate concerns for the species' survival. The Alaska breeding population is estimated to number between 35,000 and 50,000 birds (Alaskan Shorebird Group 2008; Morrison et al. 2006 in Andres et al. 2012). The oldest recorded Pacific Golden was at least 21 years, 4 months when recaptured in banding operations, and their average life span is 15 years.

▶ (Eurasian) Dotterel
Charadrius morinellus

BIOMETRICS 8–9½ inches long; wingspan: 23–25 inches; weight: 86–200 grams (3–7.1 ounces)
STRUCTURE Similar to Eurasian Golden-Plover but more compact, with a heavier chest, smaller bill, and more upright stance
STATUS Eurasian species; scarce but annual visitor and sporadic breeder in nw. Alaska; very rare along Pacific Coast, mainly in September

Barely a North American breeder, this small, colorful tundra nester breeds sporadically in the western Alaskan mountains, but most references focus only on its Old World breeding range of Scandinavia to Siberia, and at widely scattered high mountain sites from Italy to Mongolia, where birds breed on alpine as well as arctic tundra. Nevertheless, there have been 16 confirmed, and probably more, breeding records in Alaska since the first confirmed specimen taken in 1900 (Kessel 1989). There are also a handful of records (all 1st-year birds) from California in fall and a spring record from Washington State in 1934 (Howell et al. 2014). There were also three records from Iceland through 2007.

Eurasian Dotterel winters primarily in Spain, North Africa, and the Middle East. Females are more brightly colored than males, who do most of the parenting. The birds are tame to the point of "foolishness" (Hall 1960). A consequence of the friendly and trusting nature of this bird has caused the name "dotterel" in England to carry a negative connotation. The term is a contemptuous label used to describe somebody considered a doting old fool.

△ This medium-sized Old World Arctic plover breeds locally from Scandinavia to Siberia. Occurring in grassy montane tundra, the bird has bred on Alaska's Bering Sea islands and adjacent mainland and may breed sparingly on the remote Seward Peninsula. This breeding male, whose plumage is less striking than the dominant polyandrous female, was photographed in July in Sweden by Mike Danzenbaker.

▶ Lesser Sand-Plover
Charadrius mongolus

BIOMETRICS 7½–8½ inches long; wingspan: 18–23¼ inches; weight: 39–79 grams (1.3–2.8 ounces)
STRUCTURE Slightly longer-legged with a longer, heavier bill than Semipalmated Plover
STATUS Asian species; rare but regular migrant on islands off w. Alaska; very rare along Pacific Coast; accidental elsewhere

When then Secretary of State William Seward set out in 1867 to negotiate the purchase of Alaska from the Russian Empire, he had no idea how much his $7.2 million land deal would enrich the coffers of American ornithology. Many Old World breeders, including at least three shorebird species, would be added to the ranks of North American breeders. These include Lesser Sand-Plover, first collected in 1922 near Cape Prince of Wales on the Seward Peninsula of Alaska, when it was known as Mongolian Plover. This small Asiatic plover was later found nesting in 1933 when D. Bernard Bull was on a collecting expedition in Alaska for the US National Museum.

Lesser Sand-Plover breeds on barren Arctic tundra and mountainous steppes and basins up to 18,000 feet above sea level, from the Himalayas to the Chukchi Peninsula and Commander Islands in far eastern Russia, where it is a common breeder (Stejneger in Bent 1962). While the Commander Islands are geographically part of the chain of Aleutian Islands which extend off the coast of Alaska, they are still Russian territory. It is conceivable that a

Winter range for Lesser Sand-Plover includes coastal regions from South Africa and Saudi Arabia east to Taiwan and Australia, and the species is a regular but rare migrant on islands off w. Alaska, primarily May through September. This messy bird was photographed in Thailand in January.

small, scattered breeding population of Lesser Sand-Plover occurs in northern and western Alaska, but this has not been confirmed. Winter range includes coastal regions from South Africa and Saudi Arabia east to Taiwan and Australia, and the species is a regular but rare migrant on islands off w. Alaska, primarily May through September.

There are five subspecies of Lesser Sand-Plover worldwide, falling into two distinct groups, which some authorities consider separate species. The northern (*mongolus*) group differs from the southern (*altifrons*) group by its larger size, more bulbous-tipped bills, darker and colder upperparts, usually clean white rear flanks (compared with mottled on *altifrons* birds), narrower rump sides, and contrastingly dark tails.

The breeding male's rufous cowl and white throat render it unmistakable. Lesser Sand-Plover nests mainly in mountains at altitudes above tree line, with birds arriving late May to mid-June, and a typical nest includes 2–3 eggs. Intermediate- to long-distance migrants, the birds depart in small flocks from early August to September to winter in coastal areas from Africa to Australia (a huge expanse). Siberian and North American breeders migrate coastally and overland south through Asia to winter in Taiwan and Australia. All North American records belong to the northern *mongolus* group.

Strictly coastal in winter, the birds may forage alone or join other shorebird species on mudflats, sandy beaches, estuaries, and occasionally airfields in search of bivalves, marine worms, insects, and crustaceans. Lesser Sand-Plover has been recorded as a vagrant in Oregon, California, Alberta, Ontario, New Jersey, Louisiana, and Wake Island (evidence of trans-Pacific migration?).

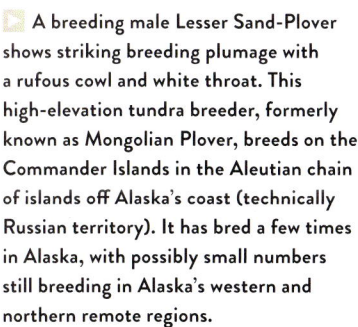

A breeding male Lesser Sand-Plover shows striking breeding plumage with a rufous cowl and white throat. This high-elevation tundra breeder, formerly known as Mongolian Plover, breeds on the Commander Islands in the Aleutian chain of islands off Alaska's coast (technically Russian territory). It has bred a few times in Alaska, with possibly small numbers still breeding in Alaska's western and northern remote regions.

▶ Snowy Plover
Charadrius nivosus

BIOMETRICS 6–6¾ inches long; wingspan: 16½–18½ inches; weight: 34–58 grams (1.2–2 ounces)

STRUCTURE Rounded and front-heavy with a short tail; compared with Piping Plover, head appears larger and blockier; body is less tapered in the rear; all-black bill is slimmer; and legs are proportionally longer

STATUS Uncommon and local; coastal populations threatened. Declining trends in most populations

----

A tiny beach pixie, Snowy Plover is one of the smallest shorebirds in North America and completely at home on high beach and dry sand. This tiny, bull-headed plover scoots across the sand with surprising speed and appears to move as if it has a series of wheels attached to its feet. (see page 97, left photo). They are often inconspicuous and hard to find on the sandy habitats their plumage resembles. These pale brown or sandy-colored shorebirds are garnished with a black or brown partial collar and a short, thin, black bill. They are proficient at snatching flies and small insects out of the air as their heads are lowered while chasing prey.

Kentish Plover, a worldwide species with which Snowy Plover was formerly grouped, is the most cosmopolitan of all plovers, with breeding populations in Europe, Asia, Africa, and the Philippines. Snowy was designated as its own species just prior to 2012 by the American Ornithological Union (AOU). It is a bird of sandy seacoasts and

◁ A winsome beach pixie, though not always "snowy," these tiny plovers fairly disappear in the shadowy depressions of the upper beach or arid portions of their inland western range. This male Snowy Plover shows plumage typical of the Great Basin or Great Plains populations and was photographed at Bolivar Flats, Texas in early April.

▽ Like something from a science fiction movie, this nonbreeding Snowy Plover is a fly's worst nightmare as it inhales tiny fly morsels while sprinting along the mud. This inland population bird was photographed in Laguna Vista, Texas in November

The western Gulf /Panhandle population of Snowy Plover is shown in this photo composite, with a female on an island beach nest off the Mississippi coast in late April, and the attending male shown nearby. This very pale resident population was formerly a separate subspecies *C. n. tenuirostris*, but recent DNA information groups it with the nominate race *C. n. nivosus*.

adjacent alkali flats as well as open flats surrounding saline lakes and braided river channels.

In North America, Snowy has several divided breeding populations of one species, *C. n nivosus*. One occurs on the West Coast from Washington State to Baja California, and another widely scattered population occurs in the interior Great Basin region of the western United States. A third exists along the Gulf Coast from Florida to s. Texas and n. Mexico. A considerable resident population also exists in central interior Mexico, and recent surveys of this and mainland Pacific Coast birds previously not surveyed substantially increased estimated population numbers.

Birds from the Gulf Coast region were formerly listed as a separate subspecies, *C. n. tenuirostris*, but genetic analysis indicates that *C. n. tenuirostris* birds are mainly restricted to Puerto Rico and Cuba, with smaller numbers on other Caribbean Islands (likely fewer than 200 total individuals (Funk et al. 2007 in Andres et al. 2012). While genetic analysis was the sole determinant of this change, Florida and upper Gulf Coast birds are substantially whiter than other populations, with breeding plumage and extent of black neck markings much less muted.

The Pacific Coast population is federally listed as Threatened in the United States, and all populations are

Nonbreeding Snowy Plovers show muted dark slashes on their upper breast sides and mostly unmarked heads and necks. The left bird is a Pacific Coast resident in February in California while the one on the right is a Great Basin/Great Plains bird on the Texas coast in September. Note the different shapes due to fluffed feathers on the left bird and an alert posture on the right one.

listed as Near Threatened by the IUCN (International Union for the Conservation of Nature), even though Snowy Plover often utilizes a polyandrous mating system that enables up to three clutches a year where a warm climate permits an extended breeding season. In these instances, females desert their mates when chicks start hatching to begin nesting with another male.

Snowy is a short-distance and partial migrant, and migration timing varies regionally. The onset of breeding varies geographically, with birds nesting as early as January in Puerto Rico; early March in California; late March in Florida; and late April to early June in the Great Basin and Great Plains. However, Kevin has seen adult Snowy Plovers and their young chicks near the Great Salt Lake in early May, suggesting a substantially earlier nest initiation date for this Great Basin location. In the Great Plains, a single brood per season is the norm.

Spring migration takes place between late February and early June. Adults typically arrive at Pacific Coast breeding sites between early March and late April, and some birds breed in more than one location per season in coastal California. Males make multiple scrapes, typically on bare ground near water, and often near some object (shell, cow pie, driftwood). Choosing her favorite site, the female deposits 2–6 buffy and lightly spattered eggs, which both males and females incubate (see page 95, upper left photo). Nests are nearly invisible on sandy beaches, where they are easily disturbed by dogs, humans, and beach vehicles. In warmer climates, adults sit on the eggs to prevent them from baking in the hot sun.

Hatching occurs in 23–28 days, and chicks depart the nest about 1–3 hours after the last egg hatches. While able to walk and peck at food, chicks are brooded and protected by adult(s) for several days. Females typically desert the chicks when they are a few days old, leaving them in the care of the male for several weeks, a scenario that harks to a high failure rate and the fate of many ground-nesting birds.

Birds nesting on the US Pacific Coast may disperse either north or south for the winter, while Florida birds are partially migratory, with some local dispersal and some birds departing the state. Birds nesting in the Great Plains head primarily to the Gulf Coast, while Great Basin nesters move primarily toward the Pacific Coast and the Gulf of California.

In winter, Snowy may form large flocks of 100+ individuals, although groups of 10–30 individuals are more likely. Snowy also likes to snuggle up against dunes, in tire

◁ A small snowball with long legs is an apt description of this very young Snowy Plover chick. Since only males tend to the chicks after a few days, mortality rate is high among chicks. Populations of Snowy Plover, other than the "Florida Gulf Coast race," have declined substantially in recent decades.

▲ Crouched on the upper beach of the Texas coast, Snowy Plovers sit tight until nearly underfoot. Birds are virtually invisible on the open beach because of the similar shading of their upperparts and sand.

tracks, or on the leeward side of driftwood. If disturbed, Snowy Plover prefers to run from perceived danger.

So well does Snowy Plover blend in with dry sand, and so tightly do birds sit, that solitary birds are nearly underfoot before they race away. Challenges faced by plovers include loss of habitat caused by beachfront development, human disturbance by recreation, beach raking that results in nest destruction and reduced food resources, and water management practices that cause flooding of nesting habitat and vegetative encroachment.

Virtually all populations of Snowy Plover appear to have declined substantially in recent decades, except for Florida, where the abundance of breeding birds appeared to be stable between 2002 and 2006 (Andres et al. 2012), although long-term datasets are generally lacking across much of the range.

Relatively new information from comprehensive interior breeding bird surveys (Thomas et al. 2012 in Andres et al. 2012) indicate a total population of 22,900, an increase of 66% from the previous estimate of 13,800 (Morrison et al. 2006 in Andres et al. 2012). For all populations of *C. n. nivosus* in North America, a proposed population of 25,900 birds exists, with a trend of apparent decline continuing. The oldest recorded Snowy Plover was 15 years, 9 months old when spotted in Oregon and identified by its band.

▽ Once dislodged, Snowy Plovers speed across the beach with such speed that they seem to be set on wheels. In flight they show relatively long wings with white undersides and a white slash with black patch on the upperwing. TEXAS, SEPTEMBER (*LEFT*); NOVEMBER (*RIGHT*)

▶ Wilson's Plover
Charadrius wilsonia

BIOMETRICS 6½–8 inches long; wingspan: 15½–19½ inches; weight: 55–70 grams (2–2.5 ounces)
STRUCTURE Front-heavy weight distribution with large, blocky head; big bill and thick neck
STATUS Locally fairly common, but strictly coastal; declining population trends suggested.

- -

This is indeed a shorebird. Intermediate in size between Semipalmated Plover and Killdeer at 6½–8 inches long, and the largest of North America's four regularly occurring small plovers, this bull-headed plover with a large bill is strictly a coastal species, where it blends in superbly with the dunes and shell-strewn habitat on high beach.

They seldom stray far from ocean coastlines and are rarely very numerous. When alert, Wilson's often assumes characteristic upright posture with steep back angle.

This species was named by George Ord for his friend Alexander Wilson, often called the father of American ornithology, after the bird was collected by Wilson in 1813 in Cape May, NJ. Birds occur on sparsely vegetated sandy beaches, adjacent salt flats, and intertidal mudflats, where they forage mostly for crabs (especially fiddler crabs, which constitute a majority of their diet).

Standing erect, Wilson's locates prey visually and pursues it with a lowered head, securing it with a large, crab-masticating bill after a short run and a neck-stretching lunge. Wilson's also has a distinctive way of running across the sand when hurried, putting its neck down and outstretching the rear body like an arrow, similar to a

◀ Found where beach meets dunes, this medium-sized, husky, large-billed Wilson's Plover has a keen sense of alertness that benefits him in the open terrain he calls home. A colorful beach in St. Augustine, Florida in April provides an attractive setting for this locally common resident.

▼ A bird of southern beaches, this most geographically restricted plover stands guard near its nest on a Mississippi coastal island in late April. The alert posture of this male reflects his protective nature while the female incubates their eggs.

Roadrunner. This unique running style allows for instant recognition, even at a distance.

Wilson's is a resident species in most locations, but a majority that nest along the Atlantic and Gulf coasts move south in winter to Central America and the Caribbean, with small numbers remaining at southern locations in s. Florida and Texas. Widespread resident populations occur in the West Indies, Caribbean, Greater and Lesser Antilles, and both coasts of northern South America.

Wilson's Plover nested in New Jersey until the 1840s (Stone 1937), and it returned as a breeder in 1935 (Boyle 2011), after which it nested irregularly for several decades before disappearing as a breeder. Its Atlantic coast range is retracting, extending from Virginia to Florida. Another migratory population extends from the central west coast of Florida to s. Texas and n. Mexico.

Three subspecies of Wilson's Plovers are recognized worldwide, with two occurring in North America. *C. w. wilsoni* breeds along the Atlantic and Gulf coasts south to Belize and the Caribbean, and *C. w. beldingi* breeds in the Gulf of California and along the Pacific Coast from central Baja California south to Peru. While typically seen in pairs outside the breeding season, loose flocks of up to 30 individuals may form, and Wilson's Plover also joins smaller beach-nesting plovers while roosting and feeding.

While apparently monogamous, Wilson's Plover sometimes breeds near other Wilson's, although birds are quite territorial near their nest site. A male makes several simple scrapes within his territory in open sandy beach close to the dunes and just above the high tide line where vegetation begins. The female then selects a favorite scrape where she deposits 3 eggs, which she both warms and shades for the next 24–25 days. The eggs are quite large compared with the body size of adults, which elicits our admiration for the female's work in producing these eggs in just a handful of days.

▶ Off to the races! Wilson's Plover runs swiftly across open spaces like a Roadrunner, with head lowered and tail on the same plane. Note the thick legs and large feet on this speedy plover. An active sight hunter, Wilson's sometimes lunges at prey.
FLORIDA, APRIL

◀ Wilson's are typically seen alone or in small groups. A male on the right displays to a prospective interested female on Bolivar Flats, Texas, in April, which hosts a good number of these coastal breeders. Wilson's is almost never found away from coastal beaches and adjacent salt flats.

▶ It seems impossible for these out-sized eggs to have been laid by this female, but here is the bird, and there are the eggs. Wilson's place their nests where the open beach starts to show some vegetation for the cover it provides, but the nest is mostly a scrape in the sand with a few shell bits and sticks nearby. TEXAS, APRIL

▼ A small barrier island near Corpus Christi, Texas in early September is the year-round home for this juvenile/1st-winter Wilson's Plover. While this bird is similar to nonbreeding adult females, the neatly arranged wing coverts and scapulars help to determine its age.

Both adults share parental duties. Typically single-brooded, many Wilson's have a second brood if the first is lost. If possible, most nests are situated close to low veg-etation, bits of driftwood, or debris. Young are able to walk 1–2 hours after hatching and depart the nest shortly thereafter, but they remain with adults who provide shade during the hottest part of the day. Otherwise, chicks find shelter among beach vegetation. Nest initiation is primar-ily mid-April in the south and early May to late June in the north. Young can fly in about 21 days.

While not globally threatened, the bird is listed as Endangered in Virginia and "a species of high concern" by the authors of the U.S. Shorebird Conservation Plan of 2000 (Sanders et al. 2012). Global population is esti-mated at 22,000 breeding birds, with 8,600 in the United States (Andres et al. 2012). Christmas Bird Counts (CBC) indicate long- and short-term declining trends, but the reliability of these data are uncertain.

▶ Common Ringed Plover
Charadrius hiaticula

BIOMETRICS 7¼–8 inches long; wingspan: 19¼–22¾ inches; weight: 39–84 grams (1.3–3 ounces)
STRUCTURE slightly chunkier body with longer rear body and wings than Semipalmated Plover; bill slightly longer and thinner than Semipalmated; toes unwebbed
STATUS rare but regular spring migrant on w. Alaska islands; accidental elsewhere; breeds in remote sections of Arctic Canada

A primarily Eurasian species, Common Ringed Plover is fairly common in its restricted North American breeding range. It barely breeds in the New World, occurring on the nw. coast of Baffin Island, e. Ellesmere Island, and Greenland, where the numbers are fairly large at around 2,000 individuals (Andres et al. 2012), substantially lower than the previous estimate of 10,000 birds (Morri-son et al. 2006). It is a regular in spring on St. Lawrence Island in the Bering Sea, where it has bred. Strangely, birds are rarely seen in North America outside their Canadian breeding range.

Common Ringed Plover is larger in size but mostly simi-lar in plumage to the more widespread Semipalmated Plo-ver. A principal distinction between these two small "ringed plovers" is the degree of webbing between the toes. Semi-palmated has partial webbing (or semipalmations) between the outer and middle toes, while Ringed Plover does not. Ringed also has a longer rear body and a thinner bill with smaller black tip; a whiter forehead that comes to a point below the eye; a dark orbital ring compared with a yellow one in Semipalmated; more extensive dark feathering on the breast and head; and a bold, white supercilium versus a restricted, muted one in Semipalmated.

Breeding male Common Ringed Plover is very similar to Semipalmated Plover, but it has a dark orbital ring compared with a yellow one in Semipalmated; a longer rear body; more extensive black feathering on the upper breast and face; and a whiter forehead that comes to a point below the eye. It also lacks the partial webbing between the outer and middle toes of Semipalmated. MAY, CYPRUS

An adult male Common Ringed Plover in worn plumage is shown here from Kenya in October with two Little Stints (Calidris minuta). Note the lack of webbing between the outer and middle toes in both photos.

Furthermore, where the ranges of the two plovers overlap on Baffin Island, the vocalizations and display patterns differ enough to suggest the two closely related species do not interbreed. Ringed Plover is accounted a "plentiful breeder" in Greenland, arriving at the end of May or beginning of June (Bent 1962). Also in this account, Dr. W. Elmer Ekblaw notes, "Few beaches are unoccupied by these noisy little birds" (Bent 1962).

Scrapes are nestled into sparsely vegetated sandy or gravelly soil, where both sexes incubate the 3–4 eggs for 22–25 days. Young leave the nest as soon as their downy fluff is dry and fledge in about 24 days. Nearctic breeders migrate across the North Atlantic to Europe in one jump, or perhaps via Iceland or Greenland, and then probably fly on to winter in w. Africa. This is unlike Semipalmated Plover, which winters coastally in North, Central, and South America. Habitat and feeding behavior of both are similar.

Recent estimates of both wintering and breeding populations suggest a total global population of 240,000–340,000 Ringed Plovers, with about 2,000 birds occurring in North America (Andres et al. 2012). Counts on the wintering grounds suggest a declining population.

Semipalmated Plover
Charadrius semipalmatus

BIOMETRICS 6¾–7¼ inches long; wingspan: 17¼–20¼ inches; weight: 31–69 grams (1.1–2.4 ounces)
STRUCTURE Chunky body but attenuated in rear with a small, rounded head; short, thick neck and stubby bill
STATUS Common in coastal areas; locally common inland; population stable

With its single black breast-band, this small, slender plover with brown upperparts somewhat resembles a demoted Killdeer, but there is nothing subordinate or submissive about this pugnacious Arctic and subarctic breeder. Semipalmated is a bold, feisty shorebird that vigorously defends prime feeding areas from other Semipalmated Plovers and, less frequently, small sandpipers. This aggression also may involve other small plovers, including Piping and Snowy (see page 109, lower photo), and is precipitated by loud, repetitive chattering, followed by an elaborate stand-off that may end with one bird landing on top of the other, a classic strategy for shorebirds who have no arms to fight with.

In migration, Semipalmated is generally tolerant of small sandpipers but quickly challenges other small plovers by charging with head lowered and tail fanned. The challenger and defender then turn and walk shoulder to shoulder, establishing for both an accepted boundary line. More often than not, the two birds face off in ritualistic fashion and then walk sideways away from each other while frequently glancing at the other plover to make sure an attack is not imminent. No other North American shorebird is as aggressive toward its own kind as Semipalmated Plover.

On wave-washed beaches, Semipalmated eschews the frenetic wave-chasing behavior of Sanderling, employing instead the calculated stop-and-go strategy of plovers, which allows birds to spot prey exposed or deposited by the receding wave. It then races forward to capture stranded prey with steps so rapid the birds appear to glide over the beach. After securing prey from the surface with their short, powerful bills, they commonly consume it on the spot, or in the case of small crustaceans, carry prey to a place away from other feeding shorebirds where the carapace is masticated in peace, without loss to pilferage.

Semipalmated is one of the most common shorebirds during migration and may occur in large, loose flocks numbering from 50 to 500 individuals. It is also one of the most widespread shorebirds in North America, breeding from northern and western Alaska east to Newfoundland and the Canadian Archipelago in Arctic, subarctic, and taiga habitats. It occurs as a migrant everywhere else on the continent other than the western mountains and Southwest region.

Part of the success of the Semipalmated Plover is attributed to the array of habitats where it breeds and forages. In subarctic regions, it nests in well-drained gravel or shale, and in northern Saskatchewan, it nests in sand dunes. In northern Quebec, it prefers rocky beaches; in Manitoba, heath-lichen tundra; and on the coasts of Hudson and James bays, shorelines, sand and gravel bars, and dry peat. In Newfoundland, it favors grassy borders of rivers and ponds and gravel runways or gravel pads laid down for oil extraction. In the Arctic tundra, Semipalmated prefers gravel shorelines of rivers and seasonal streams as well as open patches of ground in grassy tundra areas. This penchant for gravel substrate carries over through migration, where birds frequently use gravel bars to roost.

Semipalmated Plover is equally eclectic in its choice of feeding areas. In general, it favors areas that are open and wet or damp, but not covered by more than a sheen of water. Other preferred feeding locations include the edges of ponds, drying lagoons resulting from snowmelt, and, in migration, freshly plowed fields and coastal beaches. Other favored substrates include muddy sand and tidal flats exposed by a falling tide as well as wet grasslands and salt marshes.

▲ Brown upperparts on this breeding male Semipalmated Plover allow it to blend into a backdrop of mud or wet sand. Note the yellow orbital ring and partial webbing between the outer and middle toes, which helps to separate it from Common Ringed Plover. NJ, MAY

▶ As one of our most pugnacious shorebirds, these birds are touchy even in migration about other Semipalmated Plovers encroaching on "their" small feeding territories, which they defend with animated aggression displays and vocal protests. These duels usually end with one or both birds walking along invisible dividing lines. NJ, MAY

▶ With feeding over, it makes energetic sense for satiated Semipalmated Plovers to roost together for safety during migration. Many eyes see approaching danger better than two, and these roosts may contain a handful to several dozen birds. A smaller Semipalmated Sandpiper sleeps on this mud border at Jamaica Bay Wildlife Refuge in NYC in August.

▼ Semipalmated Plovers often lay their eggs in a small scrape with a few sticks on riverside gravel bars in Arctic, subarctic, and taiga regions. Four eggs are typical, but three are often seen in a re-nest. ALASKA, JUNE

Marine worms constitute the bulk of prey in coastal areas, and these are secured by a steady pull that involves the entire upright body. The birds also foot-patter to send vibrations through the damp substrate to induce worms to flinch or move, where they are easily spotted by the bird's rod-cell–packed eyes or felt by the sensitive soles of their feet (see page 19, upper left photo). Other prey include amphipods and bivalves, and away from the coast, insects and their larvae, spiders, worms, and snails.

Spring migration begins with adults departing southern wintering areas mostly in March and April, with peak numbers passing through the Gulf Coast in late April; the Pacific Coast and interior West in late April/early May; and the mid-Atlantic Coast and upper Midwest in mid to late May. Arrival on the breeding grounds takes place in early May in Alaska and British Columbia, but mostly in late May and early June in Manitoba (males before females).

▼ A Semipalmated Plover nest in the high Arctic appears as a neat table centerpiece amid attractive vegetation in ancient dunes near the Beaufort Sea, Alaska in June. The attending male's spirited broken-wing display at right evidently failed, but fortunately the transgressor was Kevin the shorebird technician, not an Arctic Fox.

Semipalmated Plovers are monogamous during the breeding season, with males guarding females as soon as they have paired and selected a nest site. Nest initiation occurs primarily mid-May in milder areas to mid-June in the Arctic. Both adults share parental duties, but females usually abandon young and the male after about 15 days. Incubation is about 23 days, and young can typically fly after 22–31 days

Most adults depart breeding areas by late July and early August (females before males), with failed breeders heading south as early as mid-June. Most birds head to the coast, but interior sites are also used, with very large concentrations at favored inland stopover sites. Juveniles migrate later than adults, usually mid to late August, with peak passage in early September.

In winter and during migration, Semipalmated forms small to large single species flocks, but birds often roost with other small plovers and sandpipers. Migration is broad-based in both spring and fall, but mostly concentrated along coastal areas. In winter, birds occur along all coasts from Virginia and Washington State south to Chile and Argentina.

Semipalmated is closely related to the Old World Common Ringed Plover, *Charadrius hiaticula*, which breeds across northern Europe, Asia, Greenland, and Baffin Island in the Canadian Archipelago, where the breeding range overlaps with Semipalmated Plover. However, the two species appear not to interbreed. The North American and world population of Semipalmated Plover is 200,000 (Andres et al. 2012), and various breeding surveys indicate a stable population trend. Migration counts increased significantly over the long term at a rate of 1.7% per year from 1974 to 2009 but have leveled off in the last decade (Andres et al. 2012).

▲ Note the crisp plumage of the juvenile bird at left (August, NY) compared with the shopworn appearance of the nonbreeding adult at right (September, Texas). The neat, crisp feather edges on the juvenile help to age this bird.

▼ While superficially similar to Wilson's Plover, Semipalmated Plover differs with a more slender, attenuated body; a smaller, orange-based bill; and a smaller, rounder head, not to mention yellow versus pink legs in this photo from Texas in early April. In flight they show relatively short, rounded wings and a compact body shape.

▶ Piping Plover
Charadrius melodus

BIOMETRICS 6¾–7¼ inches long; wingspan: 18–18¾ inches; weight: 43–63 grams (1.7–2.2 ounces)
STRUCTURE Chunky body shape with large, rounded head and large eyes; short, thick neck; stubby bill
STATUS Uncommon and local; threatened everywhere and listed as Endangered in Canada and the Great Lakes region

Bound to the narrow strip of open sand between water's edge and vegetated dunes, this small, pallid cherub with the plaintive whistled call walks a narrow line between survival and not. Now designated as federally Endangered or Threatened depending on the region, Piping Plover seemed on the brink of extinction in the late 1800s when fashionable ladies concluded that Piping Plovers made the ideal centerpiece for Victorian bonnets. Today, despite a century of recovery and intensive management, Piping Plovers are challenged or threatened by human recreational practices on the Atlantic Coast and water control practices in the interior, not to mention a host of predators.

There are three separate Piping Plover populations: an Atlantic Coast population stretching from Newfoundland to North Carolina; a northern Great Plains population stretching from Alberta to Manitoba and south to the Texas panhandle; and a smaller western Great Lakes population. The Great Lakes population had 74 pairs of Piping Plovers return to nest in 2021 (148 birds), which has steadily increased from a low of 12–17 pairs in 1981 and 108 birds in 2012 (Boissoneault 2021). A captive rearing program has shown promise here, with many fledged chicks successfully released after protective rearing in a secure facility All three populations have shown a steady increase since the 1980s due to intensive conservation efforts and nest habitat protections

Recent population trends, however, show increasing or stable numbers in both subspecies at all breeding locations (Andres et al. 2012), but the species is still considered one of high concern. The Great Lakes population is federally Endangered, and the Massachusetts, New York, and New Jersey populations are listed as Endangered by those states. All other breeding groups are listed as federally Threatened. Piping Plovers are seldom found in large numbers, except at some favored wintering or staging sites, where numbers sometimes reach 100 or more. Kevin saw and photographed a resting group of 130+ birds in early April 2005 at Bolivar Flats, Texas, a favored staging area for interior breeding birds.

While the bane of beach town officials who are challenged to balance human recreation demands with state and federal regulations relating to beach nesting bird protections, the need to monitor and protect nesting plovers is a boon to young biologists who often find entry level jobs in the conservation field as beach-nesting bird technicians. Their days are often consumed entreating dog

▶ Like pallid cherubs, Piping Plovers tread a fine line between the upper beach and water's edge, where they pick and lightly probe for food. This breeding male is foot pattering in the Atlantic Ocean surf line to cause mole crabs and other invertebrates to react to his foot motion by moving, after which he will probe the sand for a tasty meal. NJ, LATE MARCH

From fall to spring, Piping Plovers may join with other small plovers to roost in small to large flocks. Birds from the Great Plains and Great Lakes populations winter along the Gulf Coast, and this flock in early April 2005 at Bolivar Flats, Texas, numbered more than 130 Piping Plovers at this critical staging area prior to migrating north. Three darker-backed Semipalmated Plovers are seen in this group, and an alert Piping Plover at left is scanning for possible aerial predators.

owners to leash their pets and discouraging partygoers from using the protective snow fencing for beach fires, but they also provide important data on nesting successes and failures and number of offspring produced each season.

Wherever they occur, Piping Plovers prefer sparsely vegetated open beaches, alkali flats, gravel bars, and sand flats, all near water. Dredge spoils and river floodplains are also acceptable. These little round plovers hide in plain sight on sandy ocean beaches and lake shorelines, blending in perfectly with their sandy-colored backs. It takes a keen eye to spot these masters of camouflage when they are not moving about, and when not foraging, they spend their time away from the water's edge, where they sometimes crouch down in tire tracks or footprints, making them virtually invisible.

While quick on their feet, Piping Plovers don't spend as much time running around as other shorebirds and often rest quietly on upper beaches. They often place their nests just above the high tide line on ocean beaches, leaving them susceptible to nest loss from storm flooding. Birds sometimes place nests in or near tern colonies, but whether they are seeking air cover from aggressive terns or are simply reacting to a shortage of prime nest habitat is unclear. Most feeding occurs within 16 feet of the water's edge, and the birds favor a moist sandy substrate.

Piping Plovers feed using a run-stop-tilt strategy, after which they peck and probe into soft substrates for marine worms, small crustaceans (especially mole crabs in coastal areas), flies, water beetles, snails, and round-worms. They also hold one foot in front of their bodies

A very worn male Piping Plover gets ready to lower himself onto 4 huge eggs in this July nest in NJ. This late nest is surely a re-nest of a previous failed one, but it is still hard to believe these very large eggs were laid by a bird of such small size.

🔺 Can you locate the Piping Plover on its nest in this wide-open beach habitat? Birds rely on blending in with their surroundings, in this instance with their white bodies resembling shell fragments. The nest is near the center of the photo. At right, a tiny chick races to the surf line to feed among beachgoers and potential predators like gulls and terns. NJ, JULY (*LEFT*); JUNE (*RIGHT*)

and vibrate it in the sand as water passes, hoping to bring invertebrates to the surface where they can grab them with their strong, stout bills (see photo on page 105). Insect prey and their larvae are more important to the diet of interior breeders who lack the diversity of food enjoyed by coastal breeders, but despite recent positive news, recovery goals have not been achieved. Massachusetts is a leader in Piping Plover conservation and has seen the greatest increase in numbers, with 700 breeding pairs

today versus 140 breeding pairs in 1986, due to the actions of landowners and beach managers who have cooperated with the state's dynamic recovery plan.

Piping Plovers return to mid-Atlantic Coast nest locations from early March to mid-April, and to interior locations as well as New England and the Maritime Provinces from mid-April to mid-May. Monogamous pairs establish territories in open sand, gravel, or cobble beaches, often near dunes or some large object. Both sexes incubate their

🔻 Mostly solitary feeders, Piping Plovers defend their territories with aggressive displays (*left*) that may be directed toward other shorebird species as well as intruding male plovers. An adult in flight (*right*) has replaced all its wing flight feathers on the breeding grounds in July except the outer four primaries prior to migrating south in August or September. NJ, JUNE (*LEFT*); JULY (*RIGHT*)

▲ While able to run within hours of hatching, this young chick (*left*) depends on adults for protection from the sun, marauding gulls, unleashed dogs, and oblivious beach strollers. Intensely managed and studied (note the leg bands, right), this endangered shorebird faces both environmental and human-related challenges, making these young juvenile birds priceless. NJ, MAY (*LEFT*); JUNE (*RIGHT*)

3–4 eggs for 26–28 days, and within hours of hatching, chicks can walk and run. Frequent brooding is important for the next several days to maintain the body temperatures of the tiny puffballs. Otherwise, chicks forage near parents, and if fortune smiles, they can fly in 21–35 days. Young chicks that venture unaccompanied to the water's edge are susceptible to predation from gulls.

Predators comprise assorted mammals, including raccoons, coyotes, feral cats, and free-ranging dogs, crows, several gull species, Black-crowned Night Heron, and beach buggies. On Atlantic beaches, chicks are challenged to reach wave-washed feeding areas because of steady human beach traffic (see page 107, upper right photo). This distraction to adult birds results in chicks being taken by marauding ghost crabs. Storms and flooding also take a toll on eggs, with storm tides and spring nor'easters causing serious damage to nests and eggs.

Chicks fledged per pair are low based on several variables. In New Jersey in 2020, the rate was 1.29 fledglings per pair—above the long-term average of 1.09 fledglings/pair but below the federal recovery goal of 1.5 fledglings/pair. That year 103 pairs nested, a 10% decrease from 2019 and well below the 2003 peak number of 144 pairs (Heiser and Davis 2020). Heiser and Davis conclude that "New Jersey appears to have a population in flux, with sharp increases and decreases every few years."

All populations of Piping Plover are short-to-intermediate migrants. In winter, Great Plains breeders migrate to the Gulf Coast, and individuals from the Great Lakes move to South Carolina and Georgia. Atlantic Coast breeders migrate to beaches ranging from South Carolina to central Florida as well as the West Indies, Bahamas, and Cuba. Researchers recently discovered that more than one-third of Piping Plovers that breed along the

◄ The winsomeness of Piping Plover led to a precarious decline of birds during the millinery era in the late 1800s, when no fashionable lady's bonnet seemed complete without a stuffed Piping Plover centerpiece. This juvenile bird is high stepping in the Atlantic surf zone in early September in NJ.

▲ When the tide recedes, wintering shorebirds gather to feed. Here three nonbreeding Piping Plovers are joined by a Snowy Plover (front center) and four Sanderlings in early April at Bolivar Flats, Texas.

Atlantic Coast spend the winter in the Bahamas. Not a bad place to overwinter.

Atlantic Coast breeding birds have one of the direst predictions for future success. Without concerted ongoing management involving restricted beach use, protective exclosures, and predator control, the *C. m. melodus* subspecies seems unlikely to survive the century along the Atlantic coastline. Recent studies in New Jersey have shown that nest exclosures are now recognized by raccoons, foxes, and other predators as a source of food inside, and these predators burrow under the sand to get in, resulting in the discontinuance of these "protective" devices by the State of New Jersey.

Rising ocean levels seem to be the bird's biggest threat, however, as nests will certainly be washed over by ocean tides on a regular basis over the coming decades as storms increase in frequency and intensity.

The inland population, however, continues to increase, thanks to conservation efforts and research, which could be the saving grace for this fragile species. Another positive factor for its survival is the unexplained long life span compared with other small plovers, with the oldest individual recorded at least 16 years old when recaptured and rereleased in 2015 during banding operations in North Dakota. It had been previously banded in Saskatchewan in 1999.

▼ "Hey, you're not mom." A juvenile Piping Plover is discovering the pugnacious side of Semipalmated Plover, which are famously fickle about other shorebirds entering their feeding territory, even when they are hatched there. This juvenile bird learned a tough lesson in pecking order and lost a tertial feather in the squabble. NJ, AUGUST

▶ Killdeer
Charadrius vociferous

BIOMETRICS 9¼–10½ inches long; wingspan: 23½–25¼
inches; weight: 65–90 grams (2.3–3.2 ounces)
STRUCTURE Unlike other plovers, with a slender body,
long tail, small head, and thin bill
STATUS Common and familiar virtually throughout
the North American continent as a breeder, migrant,
or wintering species

It's the shorebird next door, or perhaps at your door, or nesting in your driveway, or atop your gravel roof. It is truly a shorebird you can see without going to the beach. Very few birds, and no other shorebirds, have acclimated themselves so well and so thoroughly to habitats modified for humans. From sports complexes to horse paddocks and railroad beds, Killdeer use human-disturbed habitats along with their preferred inland prairies and grasslands for nesting. They frequently construct their attractive ground nests on mowed lawns or grassy areas surrounding corporate complexes or wildlife refuge headquarters, and since they breed early in the spring, they often survive unknowing or uncaring mowers.

Killdeer also nest on flat roofs of commercial buildings, which can be problematic for the precocial chicks that need food after leaving the nest soon after hatching. Adults lead the young fluff-ball chicks to the edge of the building, or call to them from the ground, where they have to take a leap of faith by jumping 20 feet or more to the ground below. Most survive this jump, which is amazing given their tiny size and fragile nature. One set of chicks survived a jump from a seven-story building! This is the equivalent of one of us jumping off a roof more than 50 feet off the ground and surviving the fall without substantial injury. Just another amazing adaptation of this resilient group of birds.

Killdeer was given its Latin name *Charadrius vociferus* by Carl Linnaeus in 1758, an appropriate name given the loud, vocal nature of this bird. Many a time Kevin has been frustrated, but also somewhat amused, by a Killdeer giving its loud vocalization that resembles its name just as he was about to get some good photographs of nearby roosting shorebirds. This loud alarm call is a signal for other shorebirds to disperse, since danger may be right around the corner. But Killdeer often seem to give this explosive call just for fun, with no danger nearby, as if playing tricks on both Kevin and the other shorebirds. It is a high-strung bird, always on edge and always ready to sound the warning alarm call, whether needed or not.

Killdeers are very common and widespread, and they breed virtually everywhere there is appropriate habitat.

▲ As North America's most widespread and identifiable shorebird, Killdeer eschews aquatic environs in favor of dry, grassy, or even vegetatively denuded habitats. Often found singly, the birds generally avoid other shorebird species. FLORIDA, JANUARY

▽ While typically aggressive toward other Killdeer during nest season, this convocation of casual birds seem to be discussing strategy for the upcoming April breeding season in Texas. A bird in flight at right shows the distinctive orange rump and bold white wing stripe. Killdeer often resemble American Kestrel in flight.

They are denizens of dry habitats, but they are also proficient swimmers and able to swim across swift-flowing waters. While attractively dressed in warm brown colors, often with rust highlights and bold, dark, double neckbands, they are easily overlooked in their open habitats when stationary. Feeding birds run along the ground in quick spurts and stop abruptly to see if they have stirred up any insects, or just to check their progress. In flight, Killdeer resemble American Kestrel because of their comparable size, long pointed wings, and deep wing strokes.

Killdeer is the most commonly seen shorebird inland, and it breeds everywhere in the lower 48 United States except for southern Florida. It nests from se. Alaska to Newfoundland, and south through the West Indies and n. Mexico. Northern breeding birds are migratory, but a large portion of breeders in the lower 48 United States are resident, as are all the birds from the Caribbean, Bahamas, and Antilles.

Spring migration involves northbound adults departing southern wintering areas primarily between mid-February and March, with peak numbers passing through much of the continent in March and arrival at most northerly breeding sites between mid-April and mid-May. The breeding season is protracted in the south and more restricted in the north. Nest initiation is primarily early April in the south, mid-May in the north. Incubation is 24–26 days, and both adults share parental duties. One brood is typical in the north, but two or three (rarely six) broods are possible in the south. Young can fly after about 40 days.

The requisite four eggs are deposited on bare ground with sparse or no vegetation, but adults may adorn the nest afterward with attractive trinkets or nearby flowers,

as most plovers do. Prime locations are slightly elevated, and birds may situate nests near objects, fence posts, cow pies, railroad ties, and gravestones. This helps to divert attention from their nest in fully open spaces. Incubating birds seem to melt into the earth, and the double-banded breast serves to break up the plover's outline.

Approach closely and Killdeer will quietly run from the nest, trusting that the cryptically patterned eggs will go unnoticed. But if the person or predator persists, or gets too close, Killdeer break into their celebrated "broken wing" display, a vocal, histrionic activity in which birds face away from the intruder, fan their orange-based tails and rump, and drag a wing along the ground while they flutter the other, all the while crying pitifully *deer...deer...deer.*

If they are successful in distracting the intentional or unintentional intruder, they use a combination of short sprints and flights to lure the trespasser out of the area. Most shorebirds use this "broken wing" display to distract intruders or predators, but Killdeer and other plovers take the distraction display to dazzling heights. Most predation of Killdeer or their eggs and chicks occurs at nest sites. The ranks of predators include reptiles, mammals (particularly raccoons and fox), birds, automobiles, and lawn mowers.

Given vulnerable ground nests, it is understandable why many Killdeer lay their clutch atop gravel roofs, despite the challenge this imposes on flightless young. Tiny puffballs on stilt-like legs leave the nest as soon as natal down dries, and adults typically lead them to water. Chicks are able to secure prey almost immediately after leaving the nest, with parental care limited to guarding, incubating, and shading. Chicks remain with parents and siblings until fledging.

Situated in gravel parking lots, river bars, elevated railroad beds, and even flat gravel roofs, Killdeer sit tight as danger approaches and break into their celebrated distraction displays at the last moment. This nest adorned with small rocks and sticks in an abandoned lot in NJ in April shows camouflaged eggs. The bird at right is apoplectic over the presence of danger near its soon to hatch nest (May, NY), and trying its best to distract a potential predator from the eggs.

No other North American plover brandishes an orange rump or double breast-bands. Killdeer's first line of defense against imminent danger is to perform this broken wing display while scurrying away from the nest, all the while showing its bright orange rump. TEXAS, APRIL

After nesting and beginning in some areas from late June to August, northern breeders gather in small to large flocks and vacate breeding areas in most of Canada and the northern United States. Migrant flocks in late fall range from a handful of birds to more than 100 individuals, and their aerial formations are fairly tight and cohesive. Any grassland or open space is fair game for feeding and roosting during migration.

Peak migration of adults is from late August in the north to early October farther south, and for juveniles from late September/early October in the north to November farther south. They migrate by day and night to warmer climates in coastal and southern areas in late fall and winter, including the West Indies and n. South America, where they seek out grasslands, beachfronts, and prairies. Severe cold or snowfall may induce movements in late fall and early winter. Diet at all times of year consists of terrestrial invertebrates, especially earthworms, which birds capture day or night.

Killdeer do not flock with other shorebirds but may forage in the same short grass habitats and mudflats. In winter and in migration, feeding Killdeer stay well apart in places that attract numbers of birds, most notably golf courses and sod farms. They may gather in small groups, seemingly to discuss the matters of the day at these times (see page 110, lower left photo). Winter range is from coastal British Columbia, s. Nebraska, and e. Massachusetts south to w. South America from n. Chile to Venezuela.

This double-banded plover of open, lightly vegetated places has staked its claim. During the 20th century, their breeding range increased north and south as increasing numbers caused expansion of their range. Formerly relegated to sandbars, grazed prairie, coastal beaches, and burned or flooded areas, this species continues to use more terrestrial habitats than other shorebirds, but for nesting purposes, a nearby water source is preferred, even if it is only a leaking faucet.

While Killdeer numbers are impressive, with an estimated 2 million birds in North America (Andres et al. 2012), they declined 47% from 1966 to 2014 (Andres et al. 2012). However, large numbers from 2012 are higher than similar analysis done in 2001, suggesting an upward trend in numbers or more efficient data collection. Population trends are negative in the long term, and many Great Plains and western states show significant short term declines.

Two Killdeer chicks at different stages of development. Note that the more advanced bird on the right is already developing the double breast-bands and elongated profile of adults, while the left bird shows only one band. Both birds were attended by their parents, with the right bird capable of short flights. NJ, JUNE

▲ The fluffy puffballs that are Killdeer chicks are able to run and feed within hours of hatching, but they are typically attended by vigilant adults for several weeks. Three tiny chicks are brooded by one of their parents near Bunker Pond in Cape May, NJ in early May.

▶ Mountain Plover
Charadrius montanus

BIOMETRICS 8½–9½ inches long; wingspan: 21½–24 inches; weight: 90–110 grams (3.2–3.9 ounces)
STRUCTURE Somewhat chunky body shape with big head, thick neck, and large eyes; more compact than American Golden-Plover
STATUS Uncommon and declining; rarely seen near coast

Mountain Plover is an anomalous shorebird that not only shuns coastal areas but also eschews damp soils. Superbly adapted to the near desert-like environment of the High Plains and Western prairies, this dirt-colored plover seeks out intensely grazed habitats, with a partiality for prairie dog colonies, whose denuded landscape offers birds an unobstructed view of the area. This panoramic view is the plover's first line of defense against birds and animals that prey on ground-nesting birds as well as their eggs and chicks.

The first specimen was collected in 1832 near the Sweetwater River, Wyoming, by John Kirk Townsend, who presented it to John James Audubon. It was Audubon who then applied the unfortunate moniker "Rocky Mountain Plover" to this prairie specialist that never frequents mountainous areas. Its understated but attractive sandy plumage blends in perfectly with the pale short grass and desert locales that it visits, and when alarmed, it simply sits down and disappears by facing away from potential danger. Ranchers, hunters, and birders often refer to Mountain Plover as the "ghost of the prairies" because of its ability to seemingly vanish from sight into thin air.

Mountain Plover is a North American endemic that breeds in the short-grass prairies of the western Great Plains and Colorado Plateau and winters primarily in the Central Valley of California, with progressively fewer birds wintering south and east to s. Arizona, n. Mexico, and s. Texas. In winter, the birds favor plowed and burned fields or short-grass habitats, where they occur in small to medium-sized loose flocks. The name is not indicative of its habitat preference, and its only affiliation with mountains is to cross them in migration, a feat this short-distance migrant accomplishes in a single bound.

Arriving in small flocks, Mountain Plover reaches breeding grounds between early March and mid-April. Courtship and pairing occur within the flocks, with some birds paired upon arrival, and birds show a high degree of fidelity to nest sites. Males excavate multiple scrapes in flat, denuded short-grass prairie and away from hills that may mask a raptor's approach. Nest initiation is primarily mid-May, and they are often placed near an object such as a cow manure pile.

Plovers (Family Charadriidae)

▶ The inappropriately named Mountain Plover is a High Plains specialist, not a shorebird of montane habitats. Sporting the shading of winter-withered grass, this male Mountain Plover is nearly invisible except for black cap, eyes, and lores against a backdrop of native prairies where they breed, including the Pawnee Grasslands of Colorado in early July.

Specialized to thrive in dry, grazed, short-grass prairie habitats and arid plains, Mountain Plover nests are shallow scrapes typically made on bare earth and lined with lichen, grass, roots, or leaves (regularly in Prairie Dog towns and sometimes in plowed agricultural land), with the eggs often partially covered with droppings of rabbits, cattle, or other mammals. Chicks leave the nest within hours of the last egg hatching and follow adults, moving farther from the nest on successive days. Hatching success ranges from 26% to 65% and varies greatly year to year. Cold, rainy years result in greater nest failure; nevertheless, most egg and chick losses are caused by predators, not weather.

The initial response to predators is to lie flat and remain motionless. If this fails, the birds may resort to a distraction display, and when confronted by herding cattle or (in days of yore) Bison, the incubating bird will explode in a feathered frenzy in the face of the lead animal in an attempt to split the herd. In a bid to offset a nest failure rate approaching 50%, the birds adhere to the old adage "don't put all your eggs in one basket." Males and females may simultaneously brood separate clutches of 1–4 eggs each, although some pairs raise only one brood. Incubation is 28–31 days, and young can fly after 33–34 days.

Once abundant enough to be hunted for the market, the population declined 80% in the last half of the 20th century from intensive conversion of prairies to agriculture and other uses, and because of increased vegetation heights resulting from declines in grazing animals (most notably Bison, Pronghorn, and Tule Elk). In the absence of suitable heavily grazed habitat, the birds frequently resort to turned agricultural land, a gambit that often leads to nest failure and is believed chiefly responsible for the bird's

▼ "Nah, nah, you can't see me. I'm a Prairie Ghost." Were the bird to turn away from you, it would fade into the sunbaked Texas grasses. This 1st-winter bird is virtually invisible if not moving in its winter habitat in Sebastian, Texas in November.

▲ Unlike Golden-Plovers, Mountain Plover remains mostly in North America year-round, though some birds move into Mexico in winter. This nonbreeding bird sat out the winter in s. Texas, where it foraged in fallow agricultural fields in November.

mid-June onward. Adults depart the breeding grounds with juveniles in early August, having spent the summer feasting on grasshoppers, crickets, and beetles. Migrants occur through the interior West in August and September and arrive on the wintering grounds between mid-September and early November.

The world population of Mountain Plover (almost all of which breed in the United States) is now estimated at a paltry 20,000 birds, with numbers possibly still declining. Little information is available on rangewide trends, and although Breeding Bird Surveys indicate a long-term decline, they lack the precision to assess population trends. Local information suggests declines in some parts of the breeding and wintering range (Andres and Stone 2009 in Andres et al. 2012), indicating an apparent long- and short-term decline.

A proposal to list the species as federally Endangered in the United States was rejected in 2003, even though the species is on the Red Watch List of Partners in Flight, which means it is one of the most vulnerable bird species in North America. Listed as Endangered in Canada, the "prairie ghost" may indeed live up to this dubious distinction in the decades ahead.

most recent declines. Predators, however, cause most chick losses, with Kit Fox being a primary menace.

Mountain Plover is a short-distance migrant, and surviving chicks join adults in pre-migration flocks from

▽ One of Kevin's favorite shorebird photos, this juvenile/1st-winter Mountain Plover earns its name "Ghost of the Prairies" by blending in perfectly with the tilled soil in s. Texas in November. The backlit light at day's end rims the profile of this subtly beautiful, soft plover, and without the cover of his vehicle, Kevin could not have gotten close enough to take this photo of such a nervous species.

▽ A breeding adult Mountain Plover has a hard time blending in with the greenery of Colorado's Pawnee Grasslands in June after a wet spring, but their habit of standing still and turning away from potential danger still reduces their profile in open spaces.

Sandpipers and Allies (Family Scolopacidae)

This is by far the largest family group within the shorebird (wader) portion of the Order Charadriiformes, which also includes gulls, seabirds, auks, and some other smaller family groups. With 98 sandpiper species apportioned across the planet as reported by the International Ornithological Committee (Gill et al. 2021), 40+ species consistently occur in North America (36 as regular breeders) and a total of 66 sandpipers have been recorded in North America. Members of the Scolopacidae sandpiper family vastly outnumber all the other shorebird families combined. All members of this family evolved from a common ancestor, one that may have also given rise to the Jacanas.

Sandpipers comprise small to large birds, and they include the smallest shorebird in the world, the tiny Least Sandpiper at a mere 4¾–5½ inches in length, and the largest in North America, Long-billed Curlew, which from the tip of its exquisitely long bill to its abbreviated tail is 20–26 inches. Female sandpipers average larger than males, and they typically have longer bills, which accounts for the range in the measurements shown here and in the rest of the book.

Most sandpipers are slimmer than plovers, with longer legs and often thinner bills that have tactually sensitive tips. These bill tips are flexible (prehensile), and sandpipers move them up and down underground. This not only allows them to locate and grasp prey, but the motion of the flexible bill tips may startle prey and cause them to move. Shorebirds sense this movement with bill tips that may be as sensitive as our tongues and are able to capture prey with no visual help.

◀ After shorebirds bathe, they lift out of the water and flap their wings to shake off water droplets. Kevin captured this Semipalmated Sandpiper lift-up, but the photo gods allowed the eye to peek through the separated primaries, making the shot extra special. NJ, MAY

▲ A head-on flight shot of this Long-billed Curlew accentuates its very long wings and torpedo-shaped body, which is aerodynamically perfect for cutting through airspace. TEXAS, APRIL

◀ Sandpipers, like this Short-billed Dowitcher, have flexible, sensitive, movable bill tips used for searching under the mud substrate for invertebrates and insect larvae. When the moving bill tips cause worms or insect larvae to move under the surface, birds sense the motion and probe rapidly to snatch their meal without seeing it. Yet another amazing adaptation of the "Perfect Bird." FLORIDA, NOVEMBER

117

Sandpipers occupy a wide array of ecological niches, from dry tundra and grasslands to coastal or freshwater wetlands, and from forest floors to open oceans. Outside the breeding season, most are highly social and often found in mixed species flocks. However, many sandpipers are also specialized feeders, as evidenced by specific habitat demands that apportion species differently.

Mostly tactile feeders, sandpipers show an array of feeding techniques as well as mating strategies that range from monogamy to polygamy (many mates) and polyandry (sexual role reversal). Males and/or females incubate a typical clutch of 3–4 eggs and tend to young chicks, which are highly precocial, requiring from parents only protection from predators and the elements until fledging. Outside the breeding season, most occur in semiaquatic or tidal environments, and all consume invertebrate prey. Flocking offers protection against predators and may facilitate foraging success, and larger flocks afford greater protection to individuals due to strength in numbers.

Most sandpipers breed in semiaquatic habitats within boreal forest or arctic/subarctic environments, and almost all are migratory, with some species vaulting hemispheres (breeding in the Northern Hemisphere and wintering in the Southern Hemisphere). Some species, however, breed in grassland or prairie habitats (Western Willet, Upland Sandpiper, Long-billed Curlew), though usually near some water, and several others breed in forested locations (American Woodcock, Solitary and Spotted sandpipers). Several members of this family rank among the planet's long-distance champions, migrating in excess of 18,000 miles every year, with some sandpipers flying nonstop during these journeys.

Outside the breeding season, their plumage is uniformly bland, usually showing a mix of black, white, gray, and brown shading. Ranked among the planet's fastest fliers, some migrating shorebirds may exceed speeds of 100 mph, but it is the synchronized flight of massed wheeling flocks that ignites our amazement. This tight pack of birds is known as a murmuration (see page 15, upper photo). The mechanics of a murmuration is not simply a matter of one bird watching the bird next to it and reacting in kind, as has been suggested, but rather that birds in tightly packed flocks appear to watch the maneuvering of the entire flock and anticipate the next movement, like a wave pulsing through the flock. They are already calculating how they will fall in step when the wave reaches them.

One of the greatest evolutionary accomplishments of sandpipers is the ability of some species to gain weight rapidly at widely spaced migratory staging areas (hyperphagy). They are able to double their weight with energy-rich fat in several short weeks, and it is precisely this accelerated binge eating that makes long-distance migration possible. Only a few larger species, most notably Bristle-thighed Curlew, Bar-tailed Godwit, and Hudsonian Godwit, are able to lay down sufficient fat reserves to see them through migratory leaps exceeding 4,000 miles. Indeed, Bar-tailed Godwit's 7,000+ mile, 7–11 day nonstop migration from Alaska to New Zealand is the longest nonstop migration of any species on the planet.

◀ **Least Sandpiper breeds in subarctic and boreal forest ecotones (taiga) across extreme northern regions of North America, but also in coastal wetlands, bogs, sedge meadows, and sand dunes at the southern edge of its breeding range. This bird appears quite cozy in its deep, protected nest cup in Churchill, Manitoba in June.**

▶ **Juvenile Sanderling has a plumage very unlike breeding and nonbreeding adults. Bold black-and-white checkering on the back combines with soft brown markings on the upper breast sides and cheek to create a unique, attractive appearance.**

▶ Upland Sandpiper
Bartramia longicauda

BIOMETRICS 11¼–12¾ inches long (similar to Mourning Dove); wingspan: 25½–27¾ inches; weight: 97–220 grams (3.5–7.8 ounces)

STRUCTURE Small, pigeon-like head and small, straight bill; long neck and tail; upright stance

STATUS Uncommon; most numerous in the Great Plains; rare in West

----

This bug-eyed, chicken-like grassland specialist was once a common fixture on fence posts during summer months in grassland and prairie habitats across North America. Raising its wings in elegant fashion after landing, Upland elicits fondness from all who witness this salute to viewers. Formerly referred to as Upland Plover until 1973 due to its large plover-like eyes in a blank face, it was officially designated Upland Sandpiper at that time. Unlike other sandpipers, Upland avoids wetlands, instead preferring to stalk across grassland habitats with jerky steps as it jabs at grasshoppers and other insect prey.

Upland Sandpiper breeds in two distinct regions: one in central and southern Alaska to the Yukon and Northwest Territories, and the other across North America's northern tier from Alberta south to Kansas and east to New Jersey and Delaware. With its relatively small size and delicate features, it is most akin in size, shape, and habits to one of the smaller curlews, especially the Eurasian Little Curlew.

Audubon found Upland "abundant" in the prairies on "either side of the Missouri," and it is remarkable how often the word *abundant* crops up in early accounts of this species. Historically, this prairie breeder did not occur in the eastern United States, but with the felling of forests and conversion of this land to pasture (the land in New England being too rocky to be turned with a plow), the birds moved east. They established themselves in the grazed meadows that supply breeding birds with the right mix of tall- and short-grass habitats (tall vegetation for nesting, shorter grass for foraging, and short to medium-tall grass to afford cover for chicks).

Spring migrants depart South American wintering areas mostly from mid-March to early April, with peak numbers along the w. Gulf Coast in late March and early April and through the n. Great Plains and New England in early May. Alaskan breeders arrive by mid-May. Most 1-year-old birds apparently breed in their first season. Nest initiation is mostly early May in the south and late May in the north. Each pair typically lays 4 eggs in shallow scrapes on the ground, where they are subject to mammalian predators and damage imposed by large-footed grazing animals.

Adults spend equal amounts of time incubating the largish eggs that hatch in 21 days. In about one week, adults stop tending to the young, which then have to fend for themselves, including food gathering. Slightly less than half of breeding efforts are successful, with most nest mortality caused by predation. Birds have one brood per season, and birds will rarely re-nest after failure. Young can fly after about 30 days. Upland Sandpipers seem more successful where multiple pairs nest semi-colonially, but mortality among flightless young remains high.

One of the defining sounds of the American prairies is the flight call of displaying Upland Sandpipers, when the male soars upward on fluttering wings and circles its

◀ All the hallmarks of Upland Sandpiper are shown here: pigeon-like head, bug-eyed expression, long, thin neck, straight thrush-like bill, and long, sturdy legs planted in grasslands. This "grasspiper" is a common fixture in western grasslands and prairies, where it shuns wetlands for dry, grassy habitats. DRY TORTUGAS, FLORIDA, EARLY MAY

Upland Sandpiper was once limited to prairies, but Europeans felled eastern forests and transformed some into grasslands, which benefited Upland. This breeder perches on an archetypal elevated post while staking out a territory at Lakehurst Naval Air Station in New Jersey in July. You can almost hear the "wolf whistle" territorial song.

territory while giving a sputtering whistled song. Their connection with native prairie is so strong that scientists consider them an indicator of the health of the prairie, along with Sprague's Pipit and Baird's Sparrow. The absence of these three birds indicates there may be a problem with the integrity of the prairie habitat.

Its principal breeding areas are in the Dakotas, Nebraska, and Kansas (accounting for about 60% of the North American population) as well as adjacent southern portions of the Canadian prairies. It reaches its highest breeding density in native prairie habitat, but it will also nest in grazed and ungrazed pastures, in fallow agricultural fields, and sometimes in hay and other crop fields. In Alaska and northwest Canada, where Uplands are scarce, birds may nest in upland tundra, mountain meadows, and elevated ridges in wetlands and floodplains.

Ideal habitats are large areas (250+ acres) relatively free of flowering plants, though fence posts and boulders serve as perches. Fields with tilled crops like wheat and cotton do not support breeding, but the birds do well in blueberry barrens in Maine as well as in non-native grasses associated with airports and military bases, which often delay mowing until July 1 to accommodate the birds.

Most adults depart breeding areas in early to mid-July (failed breeders in late June), with peak numbers passing through the n. Great Plains in mid to late July, the central Plains in early August, and the Gulf Coast in late August. Juveniles depart mostly between late July and mid-August. Smaller numbers pass down the East Coast, peaking in late July and early August. First arrivals in South America take place in mid to late August. Most birds are gone from North America by mid-October. Upland Sandpiper

Upland Sandpipers fly circles around their breeding territories while making classic sputtering, whistled calls. Note the very long, slender wings that allow Upland to fly long distances in migration to South America with relative ease. NJ, JULY

△ Showing pale inner wings, this displaying adult is gliding over its nest territory on fluttering wings. Breeding birds perform this type of flight throughout the nesting season to survey their breeding grounds, monitor their chicks, and watch for potential intruders. NJ, JULY

winters in South America east of the Andes Mountains from Suriname and n. Brazil to n. Argentina and Uruguay, where they feast on grasshoppers and crickets, similar to their diet on the American prairies.

The numbers of Upland Sandpiper are considerably lower than the days when market gunners supplied thousands to city markets (from the late 1870s to 1890s, an estimated 50,000–60,000 birds were shipped from

Nebraska alone (Dinsmore 1994). By the time Pete set out to find Upland Sandpiper in the 1960s, the birds were breeding at only a handful of scattered grassland sites in New Jersey. Uplands were, however, a regular fall migrant in Cape May, NJ, and they might spend the day foraging in the cattle-cropped South Cape May Meadows. More often though, they would overfly Cape May, all the while making their signature *kwidy-quit* whistled call. This call is so

◁ Whether breeding in North America, wintering in the Pampas, or migrating through Texas in April, Uplands seek out grasslands. This migrant is on the hunt for grasshoppers, its favorite prey.

far reaching that one could finish a telephone conversation and rush outside in time to see the Upland come into view. Upland Sandpiper typically occurs as a solitary migrant and is rarely seen with other shorebirds in migration.

Market gunning was not the only challenge to Upland's population numbers. The eggs of Upland Sandpiper were regarded as a delicacy, comparable to the prized plover eggs of English markets. Nevertheless, in recent times, habitat loss more than gunning remains the bird's principal challenge. A time may come when people no longer hark to the "wolf whistle" call of breeding birds that so enthralled early naturalists. This would indeed be a pity.

It's a tough business being an upland ground-nesting bird, but according to North American Breeding Bird Surveys (BBS), overall population numbers of Upland Sandpiper have remained stable from 1966 to 2015, with the Great Plains breeding population being fairly common and increasing in numbers, and all other populations declining. With a BBS estimate of 350,000 birds in 2001, and a revised estimate of 1.1 million birds in 2011 (Houston and Bowen 2001, Houston et al. 2011), Andres et al. (2012) suggest in "Population estimates of North American shorebirds, 2012" a midpoint between these two estimates for a total world population of 750,000. This is a substantial increase from the 400,000 birds shown in Morrison et al. (2001). Overall, the population trend appears to be increasing. The oldest recorded Upland Sandpiper was 8 years, 11 months old and bred in New York State.

▶ Bristle-thighed Curlew
Numenius tahitiensis

BIOMETRICS 16–17½ inches long (about the same as Whimbrel); wingspan: 30–35½ inches; weight: 310–800 grams (11–28.5 ounces)
STRUCTURE Same as Whimbrel
STATUS Rare and local endemic breeder in w. Alaska; winters on South Pacific islands

- -

"They found it." The headline made news across the planet. In the summer of 1948, David Allen, son of ornithologist Arthur A. Allen, spotted a curlew in flight near the mouth of the Yukon River in Alaska. Following it to a nearby plateau, his steps became strides and the bird flushed, disclosing a nest with four eggs. This represented the first Bristle-thighed Curlew nest ever seen by non-native peoples, and one of the last North American species to have its eggs and nest described.

The eggs were first collected in Tahiti in 1785 on Captain Cook's expedition, but it took nearly two centuries for the mystery about the curlew's natal area to be solved. However, as early as the 19th century, ornithologists surmised that Alaska was the likely breeding place.

Named for a few stiff feathers on the flanks of these Whimbrel-like curlews (see photo below), the birds apportion their year between remote inland tundra breeding sites of the Yukon River and Alaska's Seward Peninsula and the tropical islands of Oceana from Hawaii to Tonga and islands

▶ Breeding on low-lying tundra near the lower Yukon River and Seward Peninsula, one of the planet's rarest shorebirds apportions its time between Alaska and islands of the South Pacific. Bristle-thighed Curlew is very Whimbrel-like except for overall buffy plumage and namesake stiff feathers near the base of the legs (shown here). ALASKA, JUNE

lying between. Yearling birds remain on wintering grounds during their first full summer, after which they return to Alaska to breed. Bristle-thighed Curlew is the only North American shorebird known to winter exclusively on an island group, with most nonbreeders residing in nw. Hawaiian Islands. Birds wintering in the southernmost islands must cross 5,000 miles of open ocean during migration.

Bristle-thighed was probably historically present year-round on the main Hawaiian Islands, but the arrival of Polynesians 1,600 years ago ended that. The birds are flightless for 2 weeks during wing molt (unlike other shorebirds, whose staggered wing molt does not result in seasonal grounding) and are vulnerable at this time to introduced mammals and other predators. While the bird's exact migration routes are presently unknown, they are believed to fly directly to and from breeding and winter destinations, with no known spring staging areas.

Birds arrive in breeding areas from early to mid-May, with southern breeders arriving first. They establish territories in interior tundra characterized by rolling hills and dominated by dwarf shrub and shrub tussock meadows. Males use aerial and ground displays to court females, and the nest typically comprises 4 eggs in a fairly deep scrape in mossy vegetation. Both parents incubate the eggs and care for the chicks for a few weeks before engaging with neighboring pairs in brood melding, after which some adults depart. This leaves the chicks in the care of a small number of adults who tend the amalgamation of young. Young can fly 21–24 days after hatching, with one brood per season.

Diet in summer consists of insects and spiders as well as fruit and other vegetation, while on the wintering grounds birds are opportunistic and omnivorous, consuming intertidal and terrestrial invertebrates, carrion, lizards, rodents, fruits, and eggs of seabirds. Bristle-thighed forages visually, walking steadily before picking, probing, or chasing prey. Some clever individuals even pilfer eggs from beneath incubating terns and frigatebirds. The eggs are either pierced with their bills or carried to a hard substrate, where the morsel is raised over their heads and forcefully slammed to the ground, thus breaking it open. Crabs and snails are decanted in similar fashion. Birds may also slam a stone or coral piece onto large seabird eggs to break them open, and such tool use is unique among shorebirds. Ghost crabs constitute an important food item when present.

Fall migration takes place between June and October, with failed breeding adults departing in mid-June, and successful breeders by late July. Adults stage along the coastal Yukon-Kuskokwim Delta between mid-June and August. Arrival at wintering sites takes place from late July to early September. Juveniles depart breeding grounds in early August, and from Alaska from mid-August to early September.

With a total breeding population estimated at 10,000 birds (Morrison et al. 2006), with 6,400 adults and 3,600 subadults, this long-distance migrant with a restricted breeding range is extremely vulnerable. Being flightless during molt also makes them vulnerable to predation by native mammals and introduced dogs and cats. The overall population has presumably declined since humans colonized the Oceana islands, but present day long- and short-term population trends are unknown.

▼ Adult (*left*) and juvenile plumages (*right*) are shown in these photos from Midway Island in the South Pacific in winter. Juveniles have less-defined streaks on the underparts, which are a richer buff color than adults, and also show larger, more buff-orange upperpart feather edges. Birds molt only once a year at wintering sites, at which time they are flightless for a short period.

▶ Whimbrel
Numenius phaeopus

BIOMETRICS   16–16¾ inches long; wingspan: 30–35½ inches; weight: 312–493 grams (11.1–17.6 ounces)
STRUCTURE   Small head and long neck with a long, decurved bill; stocky body but tapered in the rear; long legs
STATUS   Locally common along coastal areas during migration and winter; generally rare inland

Cry of the Curlew

Few birds elicit more of a primal response in humans than Whimbrel, whose series of mellow, piping whistles and soft whistled *cur-lee* calls given during migratory flights capture our attention and may result in a moment of personal introspection that transcends time. Peter Matthiessen captures this feeling perfectly in his book *The Wind Birds* when he talks about an ancient folklore of bad luck in England coinciding with the departure of curlews and plovers just before a storm.

Matthiessen wrote: "Both birds were known as harbingers of death, and in the sense that they are birds of passage, that in the wild melodies of their calls, in the breath of vast distance and bare regions that attends them, we sense intimations of our own mortality, there is justice in the legend. Yet it is not the death sign that the curlews bring, but only the memory of life, of high beauty passing swiftly, as the curlew passes, leaving us in solitude on an empty beach, with summer gone, and a wind blowing."

This relatively large curlew with a long, decurved bill manages very short turnaround times between north- and southbound migrations. These forceps-billed shorebirds snap up fiddler crabs in the marshes of mid-Atlantic states in April and May before departing for Arctic breeding grounds several thousand miles away. Before you know it, they are back in coastal marshes by late June (failed breeders) and July (mostly females) before heading for wintering territories from the southern coastal United States to South America.

But turnaround they do, raising (up to 4) chicks able to fly in 34–42 days, with both parents tending them during this time. Incubation of the eggs by both parents takes 27–28 days, and juveniles migrate south about 4 weeks after adults. The genus *Numenius*, to which Whimbrel is assigned, comes from a Greek word that means "new moon," a reference to the shape of its bill that resembles a crescent moon.

Once known as Hudsonian Curlew, Whimbrel's name has its root in the Old English "whimper-nel," perhaps a reference to the bird's stuttering call, *quiquiquiquiqui*, as rendered by David Sibley (2000). Whimbrels breed in subarctic and alpine tundra. In North America, there are two distinct breeding ranges: one in Alaska and the adjacent Canadian provinces of Yukon and Mackenzie Delta, and another on the west side of Hudson Bay in central Canada. Habitats range from dry upland heath to hummock tundra and grassy taiga bogs. When birds arrive on breeding territories, they survive on the previous summer's berries (blueberries, crowberries, and cloudberries)

▶ As shown by this Whimbrel, curlews are large-bodied shorebirds with long decurved bills that inhabit open treeless terrain. This breeding adult on the tundra near Churchill, Manitoba in June shows a distinctive striped head pattern.

before switching over to insects as these become seasonally abundant. Fall migrants also feast on berries.

Spring migration takes place from early March to early June, and arrival on the breeding grounds in Alaska occurs in early to mid-May and in Canada in late May. Nest initiation is primarily in late May in most of Alaska and early June in Canada and Alaska's North Slope. Incubation is 27–28 days, and both adults share parental duties. Young can fly after 5–6 weeks, a long time for shorebird chicks to be flightless.

Fall migration takes place between late June and late October, with failed breeders departing in mid-June and successful breeders by late July. Southbound juveniles depart breeding grounds mostly in early August. Arrival on the wintering grounds takes place mostly in late August and early September. In migration, Whimbrel makes both coastal and oceanic flights, and there are separate migration routes for western and eastern breeding populations. Some birds make nonstop flights between southern Canada and New England and South America, a distance of more than 2,500 miles.

Shorebird biologist Fletcher Smith put satellite transmitters on a handful of Whimbrels prior to 2010 and was surprised to see that his birds flew from Hudson Bay far out over the Atlantic Ocean for 1,000 miles before turning south and flying nonstop to South America. This scenario seems to be a waste of time and energy, but the Whimbrel must know something that we do not.

Dr. Smith surmised they were accessing a known north-south oceanic current in the middle of the Atlantic Ocean

▲ Breeding on moorlands and subarctic tundra, Whimbrels construct a shallow bowl-like scrape which they line with lichen and bits of vegetation. Two to four extremely large eggs are incubated by both adults, who jointly care for young. Note how carefully this large, heavy bird is sliding onto the eggs so as not to crack them open. CHURCHILL, MANITOBA, JUNE

◄ Ever vigilant, this incubating adult is on the lookout for foxes, wolves, and photographers. While generally tolerant of other breeding Whimbrel, they are dogged defenders of their airspace against aerial predators. CHURCHILL, MANITOBA, JUNE

A juvenile Whimbrel on Long Island, NY in August is migrating south for the first time. Adept at both picking and probing, the birds forage on uplands, wetlands, and sandy beaches. Note the very short bill which hasn't grown to its full length yet, and the neatly arranged wing coverts which age this bird as a juvenile. Thanks to David Speiser for this wonderful photo.

that gave them a tail wind all the way to South America. This more than made up for the migratory detour in time and energy demands that took them halfway to Europe before heading south, thus avoiding bad weather along the Atlantic coast. Benjamin Franklin noted this current during his trips to England, but we cannot help but wonder how these birds know to fly in this seemingly misdirected pattern to maximize their chances of completing their very long migration!

One celebrated Whimbrel named Machi was captured and fitted with a satellite transmitter on the coast of Virginia in August 2009. For the next two years, biologists at William and Mary College tracked Machi for 27,000 miles from Hudson Bay, Canada, to Sau Luis, Brazil, a feat that included several 2,000+ mile nonstop flights. After flying through an Atlantic Tropical Storm named Maria, Machi was shot by a hunter on Guadalupe Island on September 12, 2011, where shorebird hunting is still legal. A truly sad and tragic fact is that tens of thousands of shorebirds of several species are killed each year by "sport" hunters in the French West Indies, particularly Barbados, Martinique, and Guadalupe. Although Machi's death resulted in some promises of restrictions on this slaughter due to political pressure in Barbados, these promises have now been mostly ignored.

Barring accident, Whimbrel is a long-lived species, and Machi might have lived up to 20 years or more if not for his untimely end. While Whimbrel is not globally

Long, pointed wings and slender, tubular bodies give Whimbrel the aerodynamic perfection that allows them to migrate thousands of miles nonstop over open ocean to reach wintering grounds in South America. ALASKA, LATE APRIL

An unusual stand-off between a migrant Whimbrel and a Great Lakes nesting Piping Plover was captured by Jamie Cunningham of Ohio. Whimbrel are known to steal bird eggs from seabird colonies in winter, and this bird took interest in the Piping Plover nest on the beach. The plover showed incredible bravery by standing up to the Whimbrel but retreated when the larger bird showed menacing postures and a charge. The plover nest survived without further incident. MAY

threatened, significant declines have been noted at Atlantic staging areas. Peak numbers of migrants along the Virginia coast declined 50% between the 1990s and 2000s (Andres et al. 2012), indicating a potential significant decline in Atlantic wintering birds. And while Whimbrel were not as impacted by market gunning as American Golden-Plover and Eskimo Curlew, their populations have likely never reached pre-market gunning population levels. Egg predation and chick mortality coupled with climatic changes appear to be the principal challenges today, even more so than hunting.

Winter range extends coastally from San Francisco Bay south to Tierra del Fuego, and on the Atlantic side from coastal North Carolina to coastal southern Brazil. Whimbrels forage mostly on beaches and tidal flats where mud is the preferred substrate. Birds roost communally in shallow water or in mangroves and forage singly or in small groups, with crabs being their prey of choice. Their curved bill appears calibrated to pluck crabs from their similarly curved burrows, which the Whimbrel accesses by circling the holes as they insert their bills. Feeding mostly in the intertidal zone, birds walk or run to find food visually, then pick or probe in that area.

The current wintering New World population is estimated at 80,000 birds, with about 40,000 occurring along the entire Atlantic and Gulf coastlines, and 40,000 along the Pacific coastline, including South America. Arctic PRISM surveys over a small portion of the combined breeding ranges in North America estimated a total population of 90,781, but these numbers require more extensive ground coverage before acceptance by the scientific community (Andres et al. 2012). In conclusion, the Atlantic Coast population is apparently declining, while the trend in the Pacific Coast population is unknown.

In winter, Whimbrel often feed in coastal tidal marshes, where an abundance of prey items are easily secured. This handsome bird was in the crystal-clear Gulf waters at Fort Myers Beach, Florida, in January, where fiddler crab dinners are readily available and on the menu.

▶ Long-billed Curlew
Numenius americanus

BIOMETRICS   20–26 inches long; wingspan: 30½–39½ inches; weight: 445–792 grams (15.8–28.2 ounces)
STRUCTURE   Heavy, rounded body; long neck; extremely long, decurved bill; small head; long legs
STATUS   Locally common, especially near coastlines, but also at interior sites

In Irish folk lore, the curlew is associated with loss and lament. While the Irish focus was the Eurasian Curlew (*Numenius arquata*) and not our Long-billed Curlew, the two species are believed by some to constitute a super-species, or species complex. The keening cries of both species (*cur-lee* or *curleeuu*) are hauntingly similar and likely the foundation of the bird's dark reputation, as recounted by W. B. Yeats in his poem "He Reproves the Curlew":

> *O curlew cry no more in the air,*
> *Only to the water of the west;*
> *Because your crying brings to my mind*
> *passion-dimmed eyes and long heavy hair*
> *That was shaken over my breast.*
> *There is enough evil in the crying of the wind.*

A prophetic poem in more ways than one, although the Irish poet who lived during the late 1800s was almost certainly unaware of the slaughter of shorebirds along the Atlantic coastline of the United States that all but banished Long-billed Curlew to the western US.

Long-billed Curlew was once accounted an abundant fall migrant from New England to Florida; John James Audubon recounts seeing massed numbers going to roost near Charleston, South Carolina. And while Long-billed was described as a "common migrant in the marshes of South Sandwich of Massachusetts about 1850," by the early 20th century, Edward Howe Forbush, a prominent ornithologist, had pronounced it "very rare or accidental." He also anticipated its "approaching extinction" (Forbush 1912).

As the largest of the North American curlew species, and indeed our biggest shorebird, Long-billed Curlew is so large that it consorts with Reddish and Great Egrets and White Ibis on tidal flats in winter. It is named for its exquisitely long, forceps-like bill that may command up to a third of the length measurement and is adept at plucking and probing for its favorite foods, aquatic invertebrates.

While Long-billed is less palatable than the smaller Eskimo Curlew, its large size, flocking proclivities, and propensity to return over and over again in response to the cries of fallen flock-mates made it too tempting a target for market gunners, who had all but eradicated it along the Atlantic coast by the late 1800s.

Perhaps more damning to the curlew's numbers than market gunning was the conversion of native prairie to agricultural land. As a prairie grassland breeder, Long-billed Curlew was extirpated from Michigan, Iowa, Illinois, Minnesota, Wisconsin, and coastal Texas by 1900. Today its breeding range is limited to grasslands of the Great Plains, Great Basin, and intermountain valleys from British Columbia east to Saskatchewan and south to California, New Mexico. and Texas.

▶ Long-billed Curlew is North America's largest shorebird and among the most striking as well. Including bill, birds are 20–26 inches in length, with females larger in size and with longer bills. The very long bill on this bird indicates a female, who may weigh from 1 to 1.75 **lbs.** TEXAS, EARLY APRIL

▶ With cinnamon wings and a rich cinnamon-buff wash to the underparts, Long-billed Curlew is an attractive and memorable shorebird. Their bulk made them a preferred target for the market gunners of old, and only their remote western breeding range and migratory routes saved them from the extinction fate of Eskimo Curlew. CA, FEBRUARY

Spring migration takes place between early February and mid-May, with peak numbers passing through the Southwest from mid-March to mid-April. Arrival at most northern breeding areas occurs by mid to late April. Some nonbreeders, presumably subadults, remain at wintering sites through the summer.

For breeding, Long-billed Curlew favors flat or rolling topography with short- and mixed-grass native prairies devoid of trees. They are among the first birds to arrive on the prairie breeding grounds in spring. Paired and unpaired males establish territories from mid- to late April and make multiple nest scrapes. Nest sites, however, are chosen by females and typically situated on flat, dry, exposed upland areas (perhaps to facilitate predator detection), and often seated close to some conspicuous object (e.g., rock, cow pie). Both male and female incubate the clutch of 4 eggs, and both adults initially attend broods, with females typically departing after 2–3 weeks and before juveniles fledge 38–45 days after hatching.

Picking is the feeding technique favored by curlews during the breeding season. Coastal Long-billed Curlews spend much of their foraging time probing deep into wet sand, or probing into moist agricultural lands for worms. Consider the advantage that a long, pincer-like bill affords a bird that hunts mostly grasshoppers during the breeding season, an insect that jumps to safety when the grass it rests on is disturbed. The curlew's boarding-house reach

◀ Similar to Upland Sandpipers, when Long-billed Curlews land they raise their wings for a second or two, exposing the beautiful cinnamon-buff underwing linings. The relatively short bill and slender body compared with the females in the first two photos strongly suggest a male.
FLORIDA, FEBRUARY

△ Elegance in motion! Long-billed Curlew in flight is a picture of unsurpassed beauty. Seeing one of these birds flying by stops most birders in their tracks. TEXAS, APRIL

means it doesn't need to crowd its prey, stretch, or even lunge, just swivel and pluck. Curlews also supplement their summer diet with small vertebrates as well as the nestlings and eggs of other prairie breeders.

Long-billed Curlews are dogged defenders of their territories. Predators of all kinds are likely to encounter a squadron of mobbing adults who stream in from adjacent territories to add their protesting cries to those of the resident birds. Kevin once saw a pair of Curlews flying up to chase a Golden Eagle from their Utah nest territory in early May.

Adults gather in small flocks and depart breeding areas by midsummer, coinciding with the time drought conditions reduce insect availability. Females and failed breeders arrive at Pacific Coast wintering sites in mid-June, and along the Gulf Coast in early July. Peak numbers pass through the northern Plains from mid-July to late August, and the southern Plains and Southwest in September. Arrival at southernmost wintering areas in Central America may be as late as mid-December. Differential timing between adults and juveniles is poorly known.

Migratory flights to wintering areas in California, Texas, and Mexico are nonstop, and birds that winter in coastal areas forage on open beaches, tidal estuaries, and tidal mudflats. Away from coastal areas, birds favor moist, grassy areas or flooded agricultural lands. These active hunters, with bills that can reach more than 9 inches long in females, feed by pecking and probing, feasting mostly on terrestrial insects in summer and mollusks, crustaceans, earthworms, and benthic invertebrates in winter. Birds are typically seen in small groups but may form flocks of 50+ in choice feeding locations.

▷ Long-billed Curlew's boarding house reach has secured this crab on a grassy strip along the upper Texas coast in Galveston in September. This juvenile bird, aged by pallid whitish-buff underparts, pointed dark centers to its wing coverts, and bold dark lines on the upperparts, will throw the crab back toward its throat and ingest it whole.

Sandpipers and Allies (Family Scolopacidae)

▶ In winter along the California coast, Long-billed Curlew often forages with Marbled Godwit, which feed in similar water depths. These two species are our largest shorebirds, with Long-billed slightly larger. Note the smallest shorebirds in the world, Least Sandpipers, at the bottom of the photo. CA, FEBRUARY

The bills have evolved to aid the birds in the quest for buried prey during the breeding and nonbreeding season, especially burrowing crabs along the coast and earthworms inland. Kevin witnessed a group of nine migratory Long-billed Curlews feeding on a beach in Texas in April and noted that they inserted their bills into a crab burrow before circling the burrow while pushing the bill deeper. Since the burrow was curved, the circling motion of the bird around the hole allowed the bill to reach the bottom of the hole, thus capturing the buried crab. These birds do probe deeper than the shorebirds with which it associates in winter, most notably Marbled Godwit and Willet.

Total population numbers for Long-billed Curlew are estimated at about 140,000, with a Canadian population of 43,000 (Andres et al. 2012). Recent Breeding Bird Surveys indicate a stable population in Canada over the long and short term, and similar patterns in the United States, with a slight increase in abundance in recent years (Sauer et al. 2011 in Andres et al. 2012).

▼ While marine invertebrates are a dietary mainstay of Long-billed Curlew, insects, grasshoppers, bird eggs, and small crustaceans may also fall prey to their forceps-like bills. With the birds foraging on the coastal salt pans of south Texas in November, this scene could be a snapshot from ancient times when this species roamed these areas in much greater numbers.

▶ Bar-tailed Godwit
Limosa lopponica

BIOMETRICS 14¾–15½ inches long; wingspan: 28–32 inches; weight: 233–455 grams (8.3–16.2 ounces)
STRUCTURE Stockier than Hudsonian Godwit, with shorter neck and legs and a longer bill
STATUS Uncommon breeder in Alaska from the Yukon-Kuskokwim Delta to Point Barrow; very rare along both coasts of North America

Bar-tailed Godwit is the planet's long-distance champion, flying more than 7,000 miles nonstop over the course of 7–11 days from its Alaskan breeding grounds to winter territories in New Zealand. Yes, other birds make longer overall migrations, with Northern Wheatear's migration extending from northern Alaska to Africa, but these migrations involve stopovers, allowing birds the opportunity to rest and feed while en route. In the fall of 2022, a 5-month-old juvenile Bar-tailed left Alaska on Oct. 13 and flew nonstop for 8,435 miles over 11 days to Anson's Bay in NE Tasmania, Australia, breaking the longest nonstop flight among birds known to scientists (Readfearn 2022). This bird, known only by its satellite tag number 234864, averaged 51 miles per hour during this marathon flight that it shared with several adults.

For Bar-tailed Godwit, it is all or nothing. Most birds appear to bypass Hawaii, and it is open ocean until touchdown in New Zealand or the eastern coast of Australia. To accomplish this feat, Bar-tailed packs on layers of fat in staging areas along the west coast of Alaska for about a month in August before departing. In fact, godwits carry the greatest fat loads of any known migrant, with some birds more than doubling their normal body weight to about 600 grams (more than 1.25 pounds) in preparation for this epic journey!

Bar-taileds also increase the size and capacity of their heart and lungs in the days leading up to departure, and even reduce the size of the digestive tract as a weight-trimming strategy. Despite the rigors of migration, nearly 90,000 Bar-tailed Godwits set out on the epic flight in August, with failed breeders departing first, and adults and juveniles leaving later in large flocks. They fly in V-formations that travel at speeds up to or exceeding 45 miles per hour and at altitudes of 1,500–3,000 meters (roughly 5,000–10,000 feet) (McCaffery and Gill 2001). The nonstop journey takes up to 11 days, but typically about 7–8 days. On the return voyage (April–mid-May), birds make an interim stop at the food-rich mudflats of the Yellow Sea between the Korean Peninsula and China, making this refueling junction critical to the bird's overall survival.

Spring migration for Bar-tailed is complex and incompletely known. *L. l. baueri*, the subspecies that breeds in Alaska, begins to depart wintering areas in New Zealand and Australia as early as February, with peak departure in March and April. First arrivals at Alaskan staging areas

▶ A summer rainstorm creates sparkles on the orange breast of this male Bar-tailed Godwit on the Alaskan tundra in June. Birds from this region of Alaska (*L. l. baueri*) migrate incredibly long distances nonstop to winter in New Zealand and Australia, a distance of 7,000–10,000 miles.

take place during the last week of April, with arrival on breeding areas in early to mid-May, and in northern Alaska by early June.

Bar-tailed Godwit is both a Palearctic and Nearctic breeder, with populations of two geographically separate subspecies occurring from Scandinavia to Siberia, and in western and northern Alaska. Breeding habitat in Alaska is treeless tundra in both mountain areas and lowland coastal plains and foothills. Both adults select nest sites, which are often in dwarf shrub habitat on a slightly elevated ridge in drier locations. During the breeding season, insects and berries (especially crowberries) are important food items. Incubation is a short 20–21 days, and both adults share parental duties. Young can fly in about 30 days.

◀ Built to go the distance, Bar-tailed Godwit's football-shaped body and long scimitar-shaped wings reduce drag and increase aerodynamic efficiency. A short dark bill and rich orange underparts identifies this bird as a male. ALASKA, JUNE

▽ Spring shorebird arrivals may find Arctic breeding areas still encased in snow. A Black Turnstone and gaunt-looking male Bar-tailed Godwit were hoping for more welcoming feeding conditions in late May on the Y-K Delta of Alaska.

▲ The Winnah! Bar-tailed Godwit, the planet's migratory champion. While other shorebirds vault hemispheres in migration, in October 2022 a juvenile Bar-tailed, like the bird at right from Japan in mid-September, flew 8,435 nonstop miles from Alaska to Tasmania, a journey that took 11 days. This bird was part of a small flock, so other Bar-taileds shared the distinction of being the planet's long-distance migratory record co-holders. The bird on the left is an adult breeding female from Alaska in June showing a long bill with pink base and mostly whitish underparts with a few splotches of orange.

Adults are vigorous defenders of their territory and particularly aggressive toward other large shorebirds, but Bar-taileds often nest near species that mount aggressive attacks against predators, most notably Black-bellied Plover, Whimbrel, and Bristle-thighed Curlew. The breeding cycle from incubation to fledging is about 51 days. Adults depart soon after young fledge and gather in coastal staging areas to lay down the fat reserves they will need to fuel their nonstop migration.

Adult movements to staging areas on the Yukon-Kuskokwim Delta take place in July/August, with juveniles following afterward. All birds gather here in large, segregated flocks. Some immature birds may spend their first and second summers after hatching on nonbreeding grounds. Outside the breeding season, beginning in staging areas, Bar-tailed Godwits' principal prey is marine mollusks, crustaceans, and worms that the birds secure by picking and especially probing on soft substrates with standing water. The long bill may be inserted to the base and the head fully submerged while probing.

Birds may feed singly but appear to have greater success foraging in flocks. While the population is not threatened, habitat alteration and net trapping for food in the critical Yellow Sea staging area is cause for concern. The total number of Alaskan breeding birds is estimated at 90,000, with a trend toward declining numbers (Alaska Shorebird Group 2008; Morrison et al. 2006 in Andres et al. 2012).

▼ Bar-tailed Godwits form large flocks in winter, and they often roost with Whimbrel, a bird of similar size. This flock in Thailand in January was part of a large group of about 10,000 wintering Bar-taileds, though not the same subspecies that breeds in Alaska.

▶ Hudsonian Godwit
Limosa haemastica

BIOMETRICS 14½–16¾ inches long; wingspan: 26¾–31½ inches; weight: 196–358 grams (7–12.7 ounces)
STRUCTURE "Athletic" look with heavy chest, sleek body, and long pointed wings; long legs; long, upturned bill (O'Brien et al. 2006)
STATUS Uncommon; migrates through Great Plains in spring (sometimes large flocks), along Northeast coast in fall (mostly small groups); most fall migrants pass unnoticed offshore; Pacific population stable; e. Canadian population may be declining

- -

Once ranked among the rarest birds on the planet, this large, lance-billed shorebird has proved to be more secluded and scattered than rare, with a world population of about 77,000 birds (Andres et al. 2012). Recent work with satellite transmitter-fitted birds has shown that Hudsonian Godwits undertake incredible nonstop migratory flights from North America to South America and back again in spring, almost matching the migration of Bar-tailed Godwit. These hard to fathom migratory journeys elevate this species to a special status among biologists and birders alike, and Kevin always feels fortunate when he sees one.

One look at Hudsonian Godwit's bill imparts the understanding that it was built for probing, with feeding birds often inserting their bills up to the hilt and submerging their heads in the process. Walking along the margins of ponds with water up to their bellies, the birds stop and probe, sometimes executing multiple probes. They also wade and swim through deeper water, stopping periodically to preen.

Spring migration occurs from March to early June, though northbound movements in South America are still under study. Numbers peak in March in Brazil, with most birds gone by late April. Arrival along the Texas coast is from late March to late May, with peak concentration in mid-May. Relatively large numbers of Hudsonians are seen every spring along the upper Texas coast in May, where they feed in flooded rice fields and wet meadows. Arrival on breeding grounds in Alaska is late April to mid-May, and in the Churchill area, late May and early June.

Surprisingly little is known about this large, striking sandpiper partly because of its remote breeding and wintering areas. Breeding birds appear to require an abundance of shallow ponds and wet meadows for foraging, and scattered trees for perching (see photo on page 8). Nests are typically situated on the tops of dry hummocks, often beneath a dwarf birch, but also on elevated tussocks in sedge-tundra marsh. Surrounding vegetation serves to conceal nest, eggs, and incubating birds. The nest, constructed by males, is a simple scrape lined with vegetation.

Breeding where taiga and tundra meet, Hudsonian nests in three widely disjunct populations from western Alaska to Hudson Bay. Known breeding areas occur on the Yukon-Kuskokwim Delta and the western Kenai Peninsula in Alaska; a second location on the Mackenzie River Delta in the Northwest Territories; and a third along the south shore of Hudson Bay. Additional breeding sites have yet to be discovered but are presumed. Nest initiation probably occurs in mid-May in southern Alaska, and later at inland sites. Incubation is about 23 days, and both adults share parental duties, including brood rearing. Young can fly after about 26 days.

Insect larvae, beetles, and aquatic invertebrates are the principal prey in summer, but in migration, plant tubers (especially sago pondweed) and earthworms are important food items at inland sites. They also eat crowberries

◁ Hudsonian Godwit establish territories in marshy Arctic tundra/taiga interfaces. This breeding male appears well pleased with his corner of the planet. Kevin equates viewing one of these enigmatic birds with seeing a unicorn, but he is still hoping for that miracle. The colorful rock lichen in Churchill, Manitoba in June enhances the bird's beauty.

By August and early September, breeding Hudsonian Godwit are already winging southward, using migration routes that carry them out over the Atlantic en route to tidal estuaries in southern South America. These birds are molting into nonbreeding plumage as they fatten up at the Jamaica Bay Wildlife Refuge, NYC, in August.

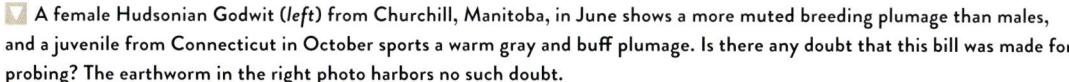
A small breeding male Hudsonian Godwit of the Alaskan subspecies is noticeably larger than nearby Short-billed dowitchers. Godwits from this region migrate south through the interior and then off the Pacific Coast to southern S. America. ALASKA, MAY

A female Hudsonian Godwit (*left*) from Churchill, Manitoba, in June shows a more muted breeding plumage than males, and a juvenile from Connecticut in October sports a warm gray and buff plumage. Is there any doubt that this bill was made for probing? The earthworm in the right photo harbors no such doubt.

in summer. In winter and during migration, including on intertidal mudflats, Hudsonian Godwit targets mostly marine worms, insects, and crustaceans. This gregarious species may mix with other probing species, most notably dowitchers, Greater Yellowlegs, and Stilt Sandpipers. Hudsonian Godwit is a compulsive "preener," often attending to only a few feathers at a time. Prior to egg laying, pairs often bathe together, and in mixed species flocks, godwits will bathe simultaneously.

Prior to long-distance migration, birds gather in small to large flocks at staging areas that include the Yukon-Kuskokwim Delta and Cook Inlet in Alaska, assorted lakes in Saskatchewan, and the shores of Hudson and James Bay. Adults begin gathering in late June and depart

for these staging areas from late June (mostly males) to mid-July (females). Juveniles and females arrive at these staging sites from mid-July to mid-August. Alaskan adults start departing for South America by late July, with most juveniles leaving by late August.

Hudson Bay nesters depart staging areas by July and August, while juveniles leave from mid-September to October. Relatively few birds are seen away from staging areas in migration, and migrating flocks may number up to several hundred birds. In spring, flocks are smaller. Nonstop flight to South America takes birds over the Maritime Provinces of Canada and New England and out over the w. Atlantic.

Insofar as much of this migration occurs offshore and at altitudes approaching 10,000 feet, it is rare for observers to see echelons of Hudsonian Godwits. Observers along the Atlantic coast are typically fortunate to see migrating Hudsonians only when offshore tropical storms or strong easterly winds ferry these powerful fliers closer to land.

During the 1980s, flocks of Hudsonian Godwits regularly numbered up to 25 birds every August at Bombay Hook NWR in Delaware, but these numbers are a memory today, with only a handful of birds present in this location during the same period, suggesting a decline in Atlantic Coast migrants (Kevin Karlson, pers. comm.).

As with many Nearctic shorebirds, Hudsonian Godwit uses an elliptical migration route, with the northbound migration west of the southbound route. In spring (March–May), northbound godwits migrate through the Great Plains, with Cheyenne Bottoms and Quivera NWRs in Kansas constituting major staging areas. Otherwise, lone godwits occasionally turn up away from staging areas, where they forage with other probing shorebirds

Hudsonian Godwit, like this breeding male from Alaska in May, undertake super-long-distance migratory flights rivaling those of their close relative Bar-tailed Godwit in distance. Instead of one long, nonstop flight, Hudsonian strings together two or three several-thousand-mile flights with extended stopovers in between.

A striking male Hudsonian Godwit feeds in Akudlik Marsh in Churchill, Manitoba, near its nest site in June while the female incubates their eggs.

but may roost with gulls, whose size they more closely approximate. In winter, virtually all Hudsonian Godwits occur in tidal mudflats of large bays, inlets, and coastal lagoons in Argentina and Chile. Important wintering areas include Tierra del Fuego and Chiloé Island, Chile.

Presumed close to extinction in the early 20th century due to market gunning of the "Ring-tailed Marlin," as Hudsonian Godwit was known to market gunners, the Pacific population is stable or slightly increasing, but likely at levels that fall short of pre-market gunning days. A recent survey of estuaries along the Pacific Coast in the vicinity of Chiloé Island, Chile (Andres et al. 2009) suggests a larger population wintering there (21,000) than previously determined from aerial surveys (14,000; Morrison et al. 2006). The southern Hudson Bay breeding population that migrates along the Atlantic Seaboard may be declining slightly based on observation records at this location from 1974 to 2009, but smaller numbers noted here may be a result of birds overflying count areas in some years. In conclusion, the Pacific population is stable and possibly increasing due to recent higher aerial counts in Chiloé Island, Chile (Andres et al. 2012), while the e. Canadian population has possibly declined according to coastal Atlantic migration counts.

Edward Howe Forbush, a noted Massachusetts ornithologist and writer, noted in 1912, "The bird undoubtedly was common here formerly in migration, particularly on Cape Cod, where it once appeared in large flocks. … Messers George M. Bubier and Lawton W. Lane report a flock of fifty birds at Ipswich on August 26, 1908, of which several were killed. This is the largest flock seen in recent years."

▶ Marbled Godwit
Limosa fedoa

BIOMETRICS 16¾–19¼ inches long; wingspan: 28–32 inches; weight: 285–454 grams (10.1–16 ounces)
STRUCTURE Bulky, football-shaped body; long legs; long bill with slight upturn to outer half
STATUS Locally common in West and South; uncommon in East; population stable

As the largest of North America's three godwit species and exceeded in weight among shorebirds only by Long-billed Curlew and the oystercatchers, this sword-billed prairie breeder occurs exclusively in North America. A cinnamon-buff plumage and strong cinnamon underwings and flight feathers give it a striking appearance that endears it to birders and casual beach strollers alike. Seeing a handful of these graceful shorebirds foraging on a shallow water mudflat is one of the sublime pleasures of birdwatching, with their large size and long, pink-based bill further enhancing their visual appeal. It is their large size combined with a very long upturned bill and cinnamon plumage that allows them to stand out from other shorebirds, especially on a wide-open mudflat.

Once more numerous in the e. United States (John James Audubon found "immense" numbers of wintering Marbled Godwits in Florida), this stately shorebird has never returned to its pre-market gunning numbers, and its breeding range has diminished in measure with the volume of native prairie habitat lost to agriculture. Breeding principally in the n. Great Plains from Alberta and sw.

▶ In winter, these large, stately, sword-billed sandpipers forage along s. North American seacoasts, probing sandy and mud substrates for aquatic invertebrates. Often found in the wave wash zone between beach strollers and surfers, Marbled Godwits may wade up to their bellies and submerge their heads while feeding. Females like the bird shown have longer bills than males. TEXAS, APRIL

▲ The superior size and cinnamon-buff color of Marbled Godwit stands out in mixed species flocks, even at a distance, as in this group at Bolivar Flats, Texas, in April. Can you identify the other shorebirds in this photo? Answer: American Avocets, Short-billed Dowitchers, Dunlin (small bird in left-center foreground), and Blue-winged Teal.

Ontario south to Montana and South Dakota, this species also has smaller isolated populations on the shores of James Bay and the Alaskan Peninsula. Marbled Godwit no longer breeds in Wisconsin, Iowa, Nebraska, or much of Minnesota as it did prior to the late 1800s.

Marbled Godwit is a relatively short- to long-distance migrant. Spring migration takes place between late March and early May, with northbound migrants leaving southern wintering areas in late March, and in mid-April to mid-May from interior locations. Arrival on prairie breeding grounds takes place from mid to late April and possibly several weeks later in James Bay and Alaska. Two subspecies are recognized: *L. f. fedora* from the Great Plains and James Bay and *L. f. beringiae* from Alaska, which are heavier and shorter winged than the inland populations.

Monogamous and territorial, the birds breed in low densities, placing their open, grass-lined nests in short, sparse upland (mostly native) grass, often well away from wetlands but concealed by taller surrounding vegetation. Both adults share parental duties, and incubation of 4 eggs lasts 23–24 days. Precocial young leave the nest 1–2 days after hatching and are attended by both parents until chicks fledge at about 30 days. As is the case with many sandpipers, females may depart before the young fledge.

While the calls of many shorebirds are distinctive, this one is idiosyncratic and mnemonic: a loud, trumpeting: *keh-WEK* or *godwit* given primarily on the breeding grounds as males circle above their large territories.

Fall migration takes place between early July and December as central interior adults form large postbreeding flocks in early July and remain through August, with numbers of birds migrating from mid-July through September. Juveniles migrate several weeks after adults, arriving on the California coast into December. Marbled Godwits are long-lived, with some birds living close to 30 years (Gatto-Trevor 2000), but the oldest recorded adult was 13 years, 4 months old when found in California, the state where it was banded, so the first statement should be taken with caution.

The flocks of long single-file strings that once navigated the Atlantic Coast from New England to Florida are gone. However, recent sightings indicate an increasing trend of wintering birds in southeastern US locations, with coastal Florida and Gulf Coast areas of Louisiana and Texas showing this species as a fairly common winter resident. Marbled Godwit is still found in winter on the flats and beaches of the West Coast from s. Washington State south to Costa Rica, and in smaller numbers along Gulf and Atlantic coastlines from New Jersey south to the Yucatan Peninsula.

Flashing cinnamon-colored underwings and brandishing its sword-like bill, Marbled Godwit is a bird that turns heads. A larger female (*left*) sports a longer bill and buffy underparts compared with a breeding male (*right*), which has more heavily barred underparts, a shorter bill, and smaller size. TEXAS, APRIL

Outside the breeding season, Marbled Godwits are highly social, often flocking and feeding with Long-billed Curlew, Whimbrel, and Willet. Foraging on mudflats and sandy beaches, the birds use their long bills to probe deeply for polychaete worms, bivalves, fish, and crabs in winter. In summer and during fall migration, birds are dependent on sago pondweed tubers, although insects and their larvae make up a majority of their diet on breeding grounds.

Birds forage primarily by walking slowly and probing like large dowitchers with bills plunged to the hilt in sand or mud and heads often submerged in standing water. On California beaches, they wait for a falling tide or reduced human traffic to gather or roost with other large shorebirds on upper beaches. Marbled Godwits sometimes feed in loose groups, unlike the slightly larger Long-billed Curlew, which is mostly a solitary hunter. They are strong fliers, with necks tucked, legs trailing, and bills thrust forward.

Breeding Bird Surveys indicate a stable population in the United States and Canada, with about 170,000 birds in the Great Plains; about 2,000 on the west coast of James Bay (2012); and about 2,000 birds on the Alaska Peninsula. The total number of Marbled Godwits in North America is estimated at about 174,000 birds (Andres et al. 2012), with a stable trend estimated for all populations. Christmas Bird Counts suggest an increasing long-term trend (Butcher and Niven 2007 in Andres et al. 2012).

At 17–19 inches long, Marbled Godwits on the Texas coast in April dwarf Short-billed Dowitchers, another probing species. Note the larger female at left with a longer bill and less heavily barred underparts than the smaller male at right.

▲ Now this is a real meal! A Black-bellied Plover steals a large lugworm from this female Marbled Godwit, which extracted the worm from deep water inaccessible to Black-bellieds. Thanks to Arthur Morris for this exciting action photo from Florida in mid-July.

▼ A nonbreeding female Marbled Godwit demonstrates that Florida sand at Fort Myers Beach in January presents no barrier to its long, probing bill. A very long bill allows godwits to access worms and invertebrates well beyond the reach of shorter-billed shorebirds.

A slender, juvenile Marbled Godwit is ready to fatten up at the Jamaica Bay Wildlife Refuge in NYC in August after recently hatching in the n. Great Plains. A lack of rich buff color on the underparts, neck, and head along with strong, dark lines of feathering and whitish spots on the back help to age this bird.

Ruddy Turnstone
Arenaria interpres

BIOMETRICS 8½–10½ inches long; wingspan: 20–22¾ inches; weight: 84–190 grams (3–6.7 ounces)
STRUCTURE Chunky body with short legs and neck; short, stout, chisel-shaped bill; low, crouched stance
STATUS Common along coasts; uncommon to rare inland; population probably declining

The high-pitched, frenetic, chattering call of Ruddy Turnstone is one of the last things a jaeger or large gull wants to hear on its Arctic breeding grounds, because it means a territorial Ruddy Turnstone is on the loose and not happy with the presence of these nest robbers in its airspace. This seemingly gentle shorebird undergoes a Dr. Jekyll to Mr. Hyde personality shift when it moves to its Arctic breeding grounds, and woe to almost any aerial predator that makes the mistake of flying into Ruddy Turnstone's nest territory.

With surprising speed (up to 47 mph) and rapid wing-beats, this torpedo-shaped shorebird flies low to the ground until it reaches the predator's airspace. Chattering angrily, it then abruptly flies directly up toward the vulnerable belly of the intruder. The predator now has two choices: fly away as quickly as possible from this screaming missile, or take a painful spearing in the belly with Ruddy's chisel-shaped bill. Most choose the former, at least those that have previously encountered this aggressive shorebird on the tundra. This behavior is in sharp contrast to the easy-going, herky-jerky foraging motion with distinctive

Harlequin-patterned Ruddy Turnstones breed across all global high Arctic regions and spend the winter on every continent but Antarctica, sometimes migrating 6,500 miles to South America. This striking breeding adult is surveying his breeding territory on the North Slope of Alaska in June.

Not a drab sandpiper at all! A breeding female Ruddy Turnstone is a poster child for complex patterns and bold colors on her upperparts. This species is one of the most aggressive defenders of Arctic breeding territories, flying low and swift across the tundra and chattering loudly before turning upward to attack aerial predators from below, against which there is no defense. After a few of these attacks, predators quickly turn and retreat after hearing the distinctive chattering alarm calls. NJ, MAY

chicken-dance head movements exhibited by this species in winter, when it rarely shows aggression to other shorebirds, but it may squabble with its own species over choice feeding holes in the sand.

This calico-patterned, cosmopolitan species breeds in high and low Arctic regions and makes long migratory jumps between bouts of gluttony to winter on rocky, pebbly, and sandy coasts and jetties in both the Northern and Southern hemispheres. Essentially omnivorous, if there is a meal to be had, Ruddy Turnstone will have it. These feathered piglets will root through kelp, pry into crevices, overturn or push aside objects to claim hidden tidbits, pillage bird eggs, pilfer scraps from feeding oystercatchers and Ospreys, and even burrow in sand up to their shoulders to exhume the buried eggs of horseshoe crabs. They are also frequent beggars for bait fish and human snacks from fishing jetties and beaches in the s. United States, even going so far as to raid active picnic lunches by walking under tables for stray food scraps.

Ruddy Turnstone's natural diet includes insects and larvae, vegetable matter, small fish, carrion, bird eggs, discarded human food, and of course, a smorgasbord of marine life, including crustaceans, mollusks, barnacles, shrimp, and starfish. Carrion and plant matter may be very important food items when they arrive in breeding areas in late May before insects are out in numbers. Thereafter, and through the summer, their diet is chiefly insects and their larvae as well as some plant material.

While horseshoe crabs are too big to flip, the viscera and trapped eggs can be grist for this omnivorous sandpiper's mill at Delaware Bay in May. Males in foreground; female in rear. A 1st-winter bird (right) feeds on mussels on a NJ jetty in December.

A group of male and female Ruddy Turnstones molting into breeding plumage wait out the high tide by lounging and feeding on a rocky jetty in St. Augustine, Florida, in early April.

There are only two global species in the genus *Arenaria* (Black and Ruddy Turnstone), both of which occur in North America. Ruddy Turnstone is one of the most northern breeding shorebirds, ranging from Arctic Alaska east across Canada to Greenland. Two subspecies are recognized: *A. i. interpres* and *A. i. morinella*. Their range is not just North America but includes wintering areas from the coasts of Great Britain and Ireland south to the coasts of se. Europe and North Africa (*A. i. interpres*).

Spring migration takes place from early March to mid-June, with most northbound adults departing wintering grounds in late March. Ruddy Turnstone is one of three sandpipers that make up a majority of the spring Delaware Bayshore food fest of horseshoe crab eggs. Arriving in early to mid-May, tens of thousands of these colorful shorebirds descend on the sandy beaches of Delaware Bay to fatten up on a bounty of high-protein crab eggs. This migratory stopover is crucial for packing on needed fat and energy reserves that enable them to make the final several-thousand-mile flight to Arctic breeding grounds.

Arriving on breeding grounds in late May to early June, they may find the tundra still encased in snow, so birds first

Ornithologist Alexander Wilson, while studying Ruddy Turnstones on New Jersey's Delaware Bayshore in the early 1800s, noted the bird's affinity for horseshoe crab eggs and named it "horse foot snipe." Photographed here in the company of Red Knots, with Laughing Gulls in the background and Semipalmated Sandpipers in front. NJ, MAY

Ruddy Turnstones choose the nest site and fashion a nest scrape to accommodate carefully arranged ellipsoid eggs. Nest sites include coastal locations near driftwood, which afford protection from the ever-present Arctic wind. Female birds do most of the incubation, and males most of the brood rearing. This female is taking 4 just-hatched chicks under her wings and belly for protection against cold Arctic weather and predators. ALASKA, JUNE

gather on the seacoasts, feeding mainly on crustaceans and mollusks. As the snow melts, birds move slightly inland, feeding in snow-free areas for edible plants and invertebrate prey (mostly insect larvae and spiders). Once the tundra is mostly free of snow, birds seek out tundra flats close to marshes, streams, and ponds that remain wet all summer, or they nest in sparsely vegetated areas close to the Beaufort Sea or other large water bodies.

Availability of food strongly influences nest site selection, and pairs show a high degree of fidelity to each other and to previously successful nest sites. After an energetic and aggressive courtship involving both ground and aerial displays, the monogamous pair reestablishes their bond. The involved courtship process and laying of eggs consumes about 10–12 days, so incubation does not typically begin until early to mid-June, with hatching from late June (Alaska) and early to mid-July (Canada and Alaska's North Slope).

Both parents share parental duties, and the male's determined defense against aerial predators is heroic, intense, and effective. Females do most of the incubation (21–23 days) and males most brood rearing, with exceptions to this norm. Family groups, including both adults, leave nest sites within 24 hours of the last egg hatching, and adults guide the young to food-rich areas along the edges of streams and ponds. Females depart before chicks fledge, but hatchlings are guarded by the male until they fledge in about 24–26 days, after which the parent/chick bond dissolves. Juveniles then leave the natal areas and form small to large flocks, occasionally with other shorebirds, most notably Red Knots.

Fall migration occurs between early July and early December, with southbound adults departing breeding grounds mostly between late July and mid-August. Failed breeders may begin to head south by early July. Their winter range is one of the most widespread of all shorebirds and mostly restricted to coastal areas. Birds that nest in w. Alaska winter from the west coast of North America to Australia and New Zealand while birds that breed in n. Canada winter on coastal areas of the United States to s. South America. Northeastern-most Canadian island and Greenland breeders winter in Europe. Juveniles migrate from mid-August to early September, or just before the first blanketing snow. Most immatures remain on wintering grounds for their first full summer.

Along rocky coastlines, Ruddy Turnstone associates with other "rockpipers" such as Purple and Rock Sandpipers, Black Turnstone, and Surfbird. It boggles the mind to see Ruddy Turnstones enduring subzero wind chills accompanied by snow and ice in ne. North America while others of their species beg for french fries in balmy tropical temperatures at Blue Waters Inn in Tobago. What determines who stays north and who goes south in winter? Your guess is as good as ours. Or could it be the french fries?

Population estimates for the three North American subspecies combined are 245,000. This estimate includes 180,000 individuals of *A. i. morinella*, which breeds in the mid to low Arctic regions of Canada; 20,000 Alaska breeding *A. i. interpres*; and 45,000 *A. i. interpres* breeding in the ne. Canadian Arctic (Andres et al. 2012). However, substantial decreases along their coastal migratory

▲ When the tide is high in late May, migrant Ruddy Turnstones in breeding plumage roost on Reed's Beach jetty in Delaware Bay, NJ. Very soon these birds will depart for a several-thousand-mile flight to high Arctic breeding grounds.

▼ After gorging on a feast of horseshoe crab eggs at Reed's Beach, NJ in late May, Ruddy Turnstones and Red Knots relocate to higher ground in a frenzy of wings.

routes, including a 77% decline along the Delaware Bayshore in May during 1988–2007, suggest a population lower than stated here (Cornell Lab of Ornithology, "Ruddy Turnstone" 2023). (Thanks to David N. Nettleship [2000] for his lucid elucidation of this species as author of *The Birds of North America* No. 537, and to the comprehensive and detailed account of this species by O'Brien et al. [2006] in *The Shorebird Guide*.)

▶ Black Turnstone
Arenaria melanocephala

BIOMETRICS 8½–10 inches long; wingspan: 20–22¾ inches; weight: 100–170 grams (3.5–6 ounces)
STRUCTURE Slightly chunkier than Ruddy Turnstone; short legs and neck; short, stout, chisel-shaped bill; crouched stance
STATUS Common along Pacific Coastlines; accidental inland; population stable to slightly increasing

Looking over the edge of the rocky Pacific coastal cliff, you first see nothing but a Black Oystercatcher, whose orange bill glows brightly against the dark backdrop of wave-splashed rocks, and a few Surfbirds whose yellow legs gleam vibrantly. Suddenly a larger than average wave breaks, and the rocky promontory fairly sprouts wings as half a dozen black-and-white–patterned birds spring from the spray and fly swiftly offshore, only to return to land a little higher up on the wave line. Against wave-blackened

rocks, Black Turnstones appear like chunks of rock until they fly.

Unlike their cosmopolitan congener Ruddy Turnstone, most Black Turnstones never leave North America, though some birds winter to s. Baja California and s. Sonora, Mexico. Indeed, this rocky coast specialist is virtually never found beyond the marine influence of the Pacific Coast, even during the 2-month period when most breed in coastal sedge meadows of w. Alaska, where many nests are within 2 kilometers of the coast. Other nests are located well inland along rivers or lakeshores, and for northernmost birds, along alpine gravel streambeds up to 2,000 feet above sea level.

Breeding from Point Hope to the Alaskan Peninsula, up to 80% of the population nests on the Yukon-Kuskokwim Delta. After breeding, the birds reside within the splash zone of the rocky Pacific Coast and nearshore islands from Kodiak Island south to Baja California. Here they utilize rocky shorelines, sand and gravel beaches, intertidal mudflats, and man-made jetties, where they forage on a wealth

▽ "This will work," a breeding-plumaged Black Turnstone is likely thinking. Late May and no snow on the tundra. Eighty percent of Black Turnstones breed on the Yukon-Kuskokwim Delta in w. Alaska, where many nests are within 2 kilometers of the coast.

▶ Spongy lichen and soft tundra make for an inviting nest site that is excavated and built by the male in Alaska in late May. Both parents incubate the eggs and care for young, and both are dogged defenders of the nest and young with swift, aggressive aerial pursuits of predators like jaegers and large gulls.

▼ So well camouflaged are Black Turnstones resting on rocky shores that it takes a wash of surf to disclose them. Two Western Willets in foreground watch the birds lifting up from a rocky shore in northern California while a very large wave gets ready to come ashore. Did you notice the surfer at the top right of this photo? CA, FEBRUARY

◀ A nonbreeding Black Turnstone feeds in seaweed and algae on rocks near La Jolla, CA, in February. Nonbreeding birds are very similar to juveniles, with distinctions linked to the more rounded shape of neatly arranged wing coverts and scapulars of juveniles. This turnstone may be a 1st-winter bird based on these criteria, but at this late date in February, definitive aging is not an easy task.

of marine organisms, especially hard-shelled invertebrates like mussels, limpets, and barnacles, which they chisel open with rapid pecks of their stout, wedge-shaped bills.

In typical turnstone fashion, the resourceful birds also flip or turn over stones, mud, algae, and other detritus found on wave-washed locations to expose prey living on or beneath these objects. They have also devised a clever way of rolling up wads of green algae to expose invertebrates hidden beneath. Birds may also "hammer" crustaceans or barnacles to loosen them or extract soft tissue. Black Turnstone is found mostly in association with Surfbird, and these two rocky coast specialists were once believed to be closely related, with both variously assigned in the 20th century to the sandpiper and the plover families.

Beginning in mid-March and continuing into mid-May, northbound spring migration occurs along the Pacific Coast or just offshore in flocks of up to several hundred birds. The principal staging area for a large portion of the population before dispersal to breeding areas is Prince William Sound, Alaska, with birds arriving in the last week of April and departing after the first week of May. Breeding areas on the Yukon- Kuskokwim Delta are reached in early May, and more northern areas from mid-May to early June.

Establishing territories in coastal tundra where snow and ice may still cover much of the ground, the birds feed on the snow's surface and in the algae exposed by melting ice. When snow has sufficiently melted, they forage extensively in wet sedge meadows, searching for aquatic and terrestrial invertebrates (especially flies) that are secured by pecking or probing. Herring eggs and mussels are also important food items.

Males precede females to breeding areas, arriving singly or in small flocks. Birds show high fidelity to breeding sites and mates of the previous season. Males begin nest-scrape displays within 5 days of arrival, or as soon as snow-free areas appear in their territories. The female selects a favorite scrape, but the nest is constructed mostly by the male and situated in a clump of last year's sedge or grass. Four eggs are incubated by both sexes for 22–24 days Adult turnstones are dogged defenders of their airspace against avian predators and have a special ire for Parasitic Jaegers (see page 26, bottom photo).

Chicks are mobile within hours of hatching and able to feed themselves, and adults lead them to areas where insects are plentiful. Brooding is shared equally between both parents, but a small percentage of females desert the brood soon after hatching, leaving males to guard chicks until they are capable of sustained flight in 25–34 days. A few adult males may accompany young to post-fledging intertidal staging areas, where they form small flocks with juvenile Ruddy Turnstones.

Southbound adults depart breeding grounds from mid-June to early September, with failed breeders departing in mid-June followed by successful breeders. Females typically depart before males. Movement timing is determined by breeding success. Juveniles fledge by mid-July and begin migration in mid-August through early September, with some continuing into November. Winter range includes coastal areas and nearshore islands from s. Alaska south to Baja California.

Population estimates for this North American Alaskan breeder remain at about 95,000 birds (Morrison et al. 2006), and CBC estimates (Butcher and Niven 2007 in Andres et al. 2012) indicate a reliable, stable to slightly increasing long-term trend.

▲ Delayed spring conditions may find birds arriving with snow still covering prime tundra nesting areas in late May. This breeding adult is searching for last year's seeds or berries or perhaps snow-chilled insects on the Y-K Delta, Alaska.

▲ Come spring's arrival on the tundra, sexual hormones of Black Turnstones are raging. No time to lose, as summers in the Arctic are ephemeral, and their bounty of insects short-lived. ALASKA, LATE MAY

▼ Where winter ranges overlap, Ruddy and Black Turnstone freely associate (*left*). Can you spot the two Ruddy Turnstones? Hint: the ones with deeper orange legs sporting a distinct collar on upper breast. A Black Turnstone in flight shows the same upperparts pattern as Ruddy Turnstone, but without the bright red colors. CALIFORNIA, FEBRUARY (*LEFT*); JULY (*RIGHT*)

▶ Red Knot
Calidris canutus

BIOMETRICS 9½–10 inches long; wingspan: 22¾–24½ inches; weight: 93–220 grams (3.3–7.6 ounces)
STRUCTURE Chunky body shape with chesty appearance; football-shaped body in flight; long wings; short legs and medium-length, mostly straight bill; horizontal stance
STATUS Locally common at stopover and wintering sites along coastlines; mostly rare inland; population has declined dramatically since 1980, especially in the subspecies *C. c. rufa* (more than 90% decline), but may be recovering as a result of conservation efforts (*see* text)

A plump, American Robin–sized sandpiper with a football-shaped body, Red Knot sports a colorful breeding plumage that includes snappy salmon-colored underparts and bold black, silver, and gold upperparts. Nonbreeding plumage is mostly dull gray with whitish underparts. It is a circumpolar breeder with five subspecies, three of which occur in North America, including one (*C. c. rufa*) which is a federally Threatened species under the Endangered Species Act and an Endangered species in a handful of states, including New Jersey.

Red Knot is a cosmopolitan species that occurs on all continents except Antarctica, and which migrates extremely long distances to winter in s. South America, Africa, and Australia, with smaller numbers in Asia. A substantial number of knots overwinter along the Gulf Coast of Florida (around 5,000 birds, Larry Niles, pers. comm. 2015), with smaller numbers wintering on the Atlantic coastline from New Jersey to Florida, and even fewer along the Gulf Coast of Texas south into Mexico.

Breeding in high Arctic tundra where few people visit or live, Red Knots spend the balance of the year in coastal marine environments, where they feed on small bivalves (especially mussels) that are swallowed whole. They crush the bivalves in the muscular part of their stomachs, known as gizzards, and they have the largest gizzards proportional to body weight of any shorebird. Another favorite food is mole crab, a soft-shell decapod crustacean that is often their food of choice if present in the coastal environs where Red Knots visit in winter. Early in the breeding season, Knots also feed on plant seeds and grass shoots, but they quickly move to invertebrates as they become available.

Knots are favorites of birders everywhere and highly sought for a variety of reasons. One is the super-long-distance migration of some knot populations from the top of the Earth to the bottom, which may involve nonstop flights of almost 3,000 miles. Another compelling attribute is their ongoing battle to survive when faced with loss of their critical stopover food source of horseshoe crab eggs in the mid-Atlantic coastal region (see *The Tragedy of Delaware Bay*, page 46).

◀ An adult Red Knot in prime breeding plumage is one of the most colorful sandpipers of all and was known to the old market gunners as "robin snipe." NJ, MAY

⬛ With a football-shaped body and long, tapered wings, Red Knot is truly designed to vault continents. Seasonal plumage and leg color differences are noteworthy in Red Knot, with strong variations between breeding (*left*, May) and nonbreeding birds (*right*, December). NJ

Red Knots are specialized to make long-distance migratory jumps between seasonally abundant food resource locations known as stopover points, where they gain weight rapidly in preparation for the next leg of their journey. At the Delaware Bay stopover point, it is not uncommon for a healthy Red Knot to almost double its body weight from about 120 grams to around 220 grams in a short 2-week period in May. This weight gain enables them to complete the final leg of their journey to Ellesmere Island and other high Arctic locations in nonstop flight, a distance of more than 2,800 miles.

If they are not able to add sufficient body fat at this critical stopover point, they will run out of body fuel and be forced to look for food in riverbeds and lake edges in the vast boreal forests of Canada, which is not an easy task. Therefore, it is paramount to maintain the health of this important migratory stopover location. Large portions of the population may be concentrated in geographically

⬛ A small group of Red Knots with Short-billed Dowitcher and Semipalmated Plover forage together on a quiet beach in St. Augustine, Florida in early April. Some knots, mostly 1st-year birds, will not molt into breeding plumage, and they won't make the long journey to Arctic breeding grounds but will remain at wintering sites or other migratory locations during the summer months.

A rush of wings captures the frenetic energy that characterizes flocks of feeding shorebirds that gather on the beaches of Delaware Bay, NJ, in May. Besides Red Knots, note the Ruddy Turnstones in the foreground and Laughing Gulls, which are also drawn to the bounty of horseshoe crab eggs.

small areas at this and other essential fueling depots, with scant backup food resources. More than 80% of the North American Red Knot population of the subspecies *C. c. rufa* may be present along the Delaware Bay in a single day, leaving them especially vulnerable to the loss of this key food resource.

"Spring" migration occurs between mid-February and mid-June as birds wintering in South America gradually move northward in February. Peak numbers occur in Brazil in late April and early May, where birds put on fat reserves that allow them to fly nonstop to staging areas along the mid-Atlantic Coast, particularly Delaware Bay. They arrive here famished and underweight, and immediately gorge themselves on lipid-rich eggs of horseshoe crabs that are deposited on the sandy beaches of the bay.

The fat accumulated at Delaware Bay must not only fuel their final flight to the high Arctic, but also sustain them until the snow melts and insects emerge, the primary summer food for adults and young. Upon arrival, some adults may consume vegetable matter in snow-free patches to tide them over. As soon as suitable habitats are snow free, males establish territories and prepare 3–5 nest scrapes. When females arrive, a male shows them by nestling into the scrape, elevating his wingtips, and kicking backward with his feet. Prime nest sites are on flat, barren, rocky, sun-drenched tundra locations, often close to vegetation and wetlands.

Males and females form traditional monogamous bonds, and both adults share parental duties, but the male does most of the incubation and most or all brood rearing, with most females leaving the natal area after the young hatch. Three or four eggs are incubated for 21–22 days, and chicks fledge in 18–20 days. Because of fluctuating seasonal snowpack amounts, there is variability in annual nest productivity. Some seasons are negatively affected by late snow melt that may result in reduced numbers of offspring. Males typically remain faithful to territories year after year.

Outside the breeding season, Red Knots are highly gregarious, feeding, roosting, and migrating in large flocks numbering in the hundreds or thousands. On the breeding grounds and in migration, they often mix with Ruddy Turnstones, but aerial migrating flocks are typically monotypic. Migrating and wintering knots use marine habitats,

High breeding plumage, anyone? These two stunning Red Knots embody flash and colorful extravagance as they navigate the shoreline in St. Augustine, Florida, in mid-April while fattening up during a migratory stopover prior to heading for the next refueling depot at Delaware Bay in May.

Not all wintering Red Knots make the energy-expensive leap to South America. This large flock of 200+ birds is feeding in the Atlantic Ocean surf in Wildwood Crest, NJ in late November, where most are gorging on an abundance of mole crabs. Some of these birds may winter in NJ or points south, including the west coast of Florida. Note the smaller Dunlin in the foreground, which are mostly hesitant to join the larger knots in deeper water.

including sandy beaches, salt marshes, lagoons, mangrove swamps, and mudflats of estuaries and bays that contain an abundance of invertebrate prey. Other habitats that might harbor knots include peat banks (remnants of ancient forests on the seashore, exposed by erosion), salt ponds, eelgrass beds, and Brazilian *restingas* (coastal spits).

Fall migration takes place between early to mid-July and mid-November, with southbound migration of adults peaking in early to mid-July. The fall migration of Red Knots that breed in Nearctic Canada (*C. c. rufa*) begins with birds staging on the coasts of Hudson and James Bays and the Atlantic Provinces of Canada. Some birds engage in long overwater flights to the northern coast of South America, while others stop along the Atlantic Coast to gorge on mollusks and one of their favorite protein-rich food sources in the receding wave zone, mole crabs. By mid to late August, these birds make the flight to their wintering grounds in Argentina.

▲ As the species name *canutus* (which means "daring" in Middle German implies), Red Knot's life is bounded by the tides. A nonbreeding bird in NJ in January appears much plumper (probably twice the weight) than the slender, crisply marked juvenile at right, which just flew several thousand miles after hatching from the Arctic to a shoreline in Stone Harbor, NJ, in September, where a plentiful mole crab feast awaits.

The population that breeds in Alaska (*C. c. roselaari*) apparently disperses to a number of sites along southern coasts of the United States, California, Central America, and perhaps n. South America, arriving by early to mid-July. Some of these birds arrive on the mid-Atlantic Coast and may remain well into September, after which most gradually move south. Juveniles head straight to the coast after spending several weeks feeding near the breeding grounds. They first arrive on the mid-Atlantic and mid-Pacific coasts of North America by mid to late August, with peak passage in September and early October. Arrival on the wintering grounds takes place between October and December, but small numbers overwinter on s. Pacific and Atlantic coastlines, including s. California.

Populations of North American Red Knot breeders have sharply declined, with a drop of more than 50% in wintering numbers in South America from the mid-1980s to 2003. The estimated population of all three North American subspecies is about 139,000 breeding birds, and the trends in the subspecies *C. c. rufa* indicate a significant decline in all three wintering populations (Andres et al. 2012).

The oldest recorded Red Knot is the famous Moonbird, also known as B95, banded in 1995 in Argentina by biologist Patricia González. B95 was resighted a handful of times until the final sighting along the Delaware Bayshore in 2016, making him at least 21 years old! His total estimated mileage during his lifetime migration was more than 400,000 miles, a distance equal to traveling to the moon and halfway back again, thus the name "Moonbird." A 2012 book titled *Moonbird: A Year on the Wind with the Great Survivor B95*, by Phillip Hoose, tells the tale of this bird and his amazing journeys against all odds (see expanded story on page 40).

▶ Surfbird
Aphriza vigata

BIOMETRICS 9½–10½ inches long; wingspan: 24¾–27¼ inches; weight: 133–230 grams (4.7–8.2 ounces)
STRUCTURE Chunky body shape with short legs and neck; short, stout bill
STATUS Locally common along rocky Pacific coastlines; accidental inland; population stable

-----

This fairly large, plump, lead-gray shorebird with a white belly is a familiar sight from midsummer to spring along the Pacific coastlines of the Americas. Often seen in the company of Black Turnstone, it scours the slick, surf-washed rocks for intertidal invertebrates, especially barnacles, snails, and bivalve mollusks. Even its breeding plumage is mostly nondescript, though heavily patterned, with a smattering of rust feathers on the upperparts, and numerous bold, black chevrons on the underparts.

For up to 11 months of the year, this sturdy "rockpiper" is a denizen of rocky coastlines from Kodiak Island, Alaska, to s. Chile, a linear distance of some 10,900 miles, giving Surfbird the longest and narrowest nonbreeding range of any native North American bird, and one of the greatest latitudinal winter ranges of any bird species. However, for several months of the year, Surfbird disappears from the ken of science. To where? Never-never Land? No, more aptly to Shangri-La.

It seems that these coastal birds established a breeding foothold on the rocky alpine tundra in the mountains of Alaska and Yukon interior, exactly where Native Americans said they would be. But it wasn't until 1921 when

It's a plover; no, it's a sandpiper! It is, in fact, a Surfbird, a highly specialized shorebird that occurs within the wave zone of rocky Pacific coastlines from Alaska to the Strait of Magellan for up to 11 months of the year. In summer they retreat to breed in high-elevation tundra habitats in Alaska and Yukon Territory. This attractive breeding bird was photographed in Homer, Alaska in early May, where a flock of Surfbirds fed on the ample barnacles.

For 8 months of the year, Surfbirds are seen in this lead-gray plumage with white underparts. Only the yellow legs and bill-base break up the drab but engaging appearance that blends in perfectly with the rocky shorelines they inhabit. CA, November

biologist Olaus J. Murie happened upon two adults and a downy chick on the slopes of Mt. McKinley, Alaska, that the mystery of Surfbird's breeding location was uncovered. In 1927, Joseph Dixon also found a nest near Mt. McKinley. It all makes perfect sense from the standpoint of breeding success, with remote locations discouraging visitation by humans and potential predators.

Mammalian predation is a major cause of nest failure among ground-nesting shorebirds, and the sterile, high-elevation slopes of northern mountain peaks are hardly worth the effort of marauding wolves, foxes, and wolverines. Even the tireless cruising Long-tailed Jaeger eschews these higher altitudes and mostly hunts lemmings and voles in the rich river valleys below.

Egg-hungry Arctic Ground Squirrels and the inopportunely placed hoofs of Dall Sheep appear to be the principal threats to the eggs and young of Surfbird. Even Gyrfalcons don't constitute an immediate hazard to juvenile Surfbirds before they fledge, after which they become fair game for this formidable aerial predator.

Spring migration takes place between mid-February and early June, with northbound adults departing southernmost wintering areas by early March. Most of the planet's Surfbirds stage in early May at Prince William Sound,

Spending most of their lives within the sound of the surf, these migrating adult Surfbirds in Kachemak Bay, Alaska, in May are catching some shut-eye while nearing the finish line in nearby high-mountain tundra habitats.

where birds lay down the layers of fat they will need to see them through the insect-impoverished first days of courtship and mating on the high-elevation breeding grounds. Unlike several other Arctic shorebirds, Surfbird is not known to forage on last year's berries to tide them over.

More than 50,000 Surfbirds, nearly the entire population of around 70,000 birds (Morrison et al. 2006), have been tallied on Montague Island in Prince William Sound in early May (Senner and McCaffery 1997). Arrival at breeding areas takes place between early May and early June. Small numbers, especially 1-year-old birds, remain at wintering areas throughout the breeding season.

Surfbird nests are situated near the summit and upper slopes of mountains on vegetated and non-vegetated dry alpine tundra, and most often on north- or west-facing slopes. The nest is a shallow depression lined with lichen,

into which the 4 lichen-patterned eggs are deposited. The incubating bird's response to an approaching sheep or biologist is to explode into the face of the transgressor in a feathered frenzy guaranteed to halt further progress and hopefully elicit a course correction.

Nest initiation is primarily late May and early June, with egg dates from mid-May to early July. Incubation of 22–24 days is probably done by the male alone since it is unknown whether females incubate the eggs. Broods are tended by both adults, but mostly by males. Fledging period is unknown, as is much of the bird's breeding biology, mostly because of the remoteness of breeding areas.

Outside the breeding season, the doughty sandpiper with a plover-like bill maneuvers alone or in small groups within the splash zone of the rocky Pacific coasts (rarely sandbars), where they may join other "rockpipers." The

Synchrony in motion! The short breeding season over, adult Surfbirds off the California coast in July (*left*) are transitioning to nonbreeding plumage. The bird on the right in breeding plumage was photographed by Kevin in Homer, Alaska in May.

short, sturdy bill of Surfbird enables them to pull barnacles and mussels off rocks with a leveraging sideways tug and swallow them whole, where the bird's industrial-strength gizzard and stomach pulverize them. Lacking the hooked bill of Black Turnstone, they rarely eat limpets.

Fall migration takes place between late June and early December, with southbound adults departing breeding areas mostly between mid-July and early August. Failed breeders and nonbreeders may begin to head south by late June. Most juveniles depart the tundra by late July or early August and gather in small flocks at nearby coastal areas. Passage of juveniles occurs about a month later than adults.

No new information is available to change the global population number of 70,000 Surfbirds in North America (Morrison et al. 2006), but Christmas Bird Counts indicate a reliable, stable to slightly increasing long-term trend (Butcher and Niven 2007 in Andres et al. 2012). While the overall population appears stable, the mass concentration of Surfbirds in spring lies in a major petroleum corridor, which is reason for concern, if not alarm.

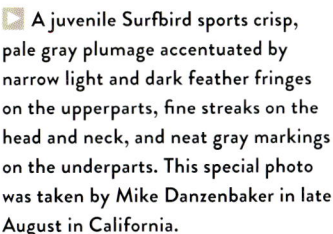 Although there is an abundance of barnacles on these rocks in Homer, Alaska, in May, these two Surfbirds are squabbling over a small feeding area while their neighbors look on with indifference.

A juvenile Surfbird sports crisp, pale gray plumage accentuated by narrow light and dark feather fringes on the upperparts, fine streaks on the head and neck, and neat gray markings on the underparts. This special photo was taken by Mike Danzenbaker in late August in California.

▶ Ruff
Calidris pugnax

BIOMETRICS 8–12 inches long, with females at the low end; wingspan: 19¼–23¼ inches; weight: 70–268 grams (2.5–9.5 ounces, with females at low end)

STRUCTURE Chunky, rounded body shape; longish legs and neck; small head with short, slightly drooped bill; hunches over when feeding (O'Brien et al. 2006). Females are smaller and slenderer overall, with shorter neck and legs and thinner bills

STATUS Rare but regular migrant in North America, mostly along Atlantic and Pacific coasts (less along Gulf Coast), upper Midwest, and w. Alaska; rare in winter in California. Population status has declined dramatically, but mainly in temperate-breeding birds. Ruff is one of the most abundant Eurasian shorebirds.

Ruff has a distinctive gravy boat appearance, with a small head, medium-length bill, longish neck, and pot-bellied body. Males are somewhat large, about the size of Black-bellied Plover, while the smaller females (reeves) are close to Pectoral Sandpiper in size. Males are famous for their gaudy breeding plumage that includes variously colored head dresses of rust, black, and white, and for their animated lek displays, when they raise their head and neck feathers in showy fashion to outcompete nearby males to win the favor of nearby females.

While the inclusion of this Old World species in a treatment of North American shorebirds is contestable, it is a rare but regular migrant in Alaska, along Atlantic and Pacific coasts, and through the Midwest, and it has bred at least once in Alaska. Though seemingly not as common in North America as they were in the 1970s and '80s, the birds were at one time annual and found in good numbers in places like Pedricktown, NJ and Thompson's Beach, NJ, where in early May up to a dozen birds might be in view at one time.

Named for the feather duster array of neck feathers donned by males during the breeding season, Ruffs are celebrated for their polychromatic plumages, with the considerably smaller females having a mostly muted plumage with a mix of buff and brown feathers.

The polygamous male exhibits dramatic "grouse-like" lek courtship behaviors to facilitate copulation with as many females as he can attract to his territory, with animated displays involving the colorful headdresses. After copulation, males have nothing to do with nest building or parental duties, leaving females to fend for themselves during incubation and brood rearing. Breeding takes place both in moist Arctic habitats and in Old World marshlands and grasslands in temperate zones.

In spring (March to mid-May), migrating North American Ruffs are typically found among Greater Yellowlegs in tidal estuaries with lush but broken habitat. Flooded corn-fields are also favored haunts. Later migrating birds, especially reeves, are more often in the company of Pectoral Sandpipers in grassy, freshwater habitats.

Some biologists speculate that Ruff may breed in Arctic or subarctic regions of Canada, which is a distinct

◀ Like something from outer space, this hormonally charged male Ruff is showing off his finery in Alaska in June. As a lekking species, males attract females to their often-adjoining territories by raising their showy feathers while strutting around, and fights often occur between competing males. Though this species has been found only once breeding in Alaska, this bird is doing its best to change that number.

This Ruff is not sporting typical breeding plumage in mid-May in NJ, but is showing plumage of the "faeder form" of Ruff, which resembles female Ruff's (reeve) plumage but with more black markings on the breast. The word *faeder* comes from the Old English word for "father." While the Latin name for Ruff translates as "fight-loving fighter," this form of Ruff forgoes the competitive displaying and occasional fighting between males and instead prefers to blend in with the females, occasionally mating with them unnoticed. Unusual but true! A smaller breeding-plumaged male Ruddy Turnstone is seen nearby.

possibility. Another scenario suggests that migrating East Coast birds may simply be individuals that cross over from w. Africa after catching trade winds to make landfall in the New World. Once here, they join Greater Yellowlegs on their northward migration. Where do they go from there? No answer, but they do not breed in Iceland or Greenland, even though the breeding range for this abundant Old World breeder spans Arctic and subarctic regions from Britain and Scandinavia to eastern Siberia.

Ruff numbers have declined in the last 50 years, but much of that decline occurred in birds that historically bred in the marshes of Europe. Breeding Ruffs have declined in all habitats across temperate Eurasia, with about 98% of Ruffs now breeding only in Arctic habitats. Despite these declines, it is still one of the most abundant shorebirds in Eurasia, with a population estimate of 2.3–2.8 million birds (Zöckler 2002). Wintering numbers are consistent with this breeding population estimate.

A displaying Ruff surveys the vast tundra in Alaska in mid-June hoping to entice a few females to his lek. With only one confirmed breeding record in Alaska, his hope for mating may not come to fruition, but the tundra is vast and mostly without people to document breeding or courtship activity, so who knows what might happen or has happened? Kevin once witnessed four Ruffs displaying near Prudhoe Bay, AK in June 1992, but no nests were found.

▶ Stilt Sandpiper
Calidris himantopus

BIOMETRICS 8–9¼ inches long; wingspan: 17¼–18¾ inches; weight: 50–70 grams (1.7–2.5 ounces)
STRUCTURE Small head with long neck and legs; fine-tipped bill droops near the tip; odd, puffy rear crown; slender, tubular body shape; longer legged and shorter billed than dowitchers; shorter legged and longer billed than Lesser Yellowlegs
STATUS Fairly common migrant through Great Plains in spring and fall, and along the Atlantic Coast in fall; uncommon in winter in s. Florida and Texas coast; scarce elsewhere

It's a dowitcher! It's a Lesser Yellowlegs! No, it's neither, although Stilt Sandpiper has qualities that remind one of both these species. It is structured somewhat like Lesser Yellowlegs but with a longer bill that droops near the tip, not straight as in Lesser, and has much shorter wings. It also has pale yellow to greenish-yellow legs that may be mostly or fully submerged when feeding, not bright yellow like Lesser Yellowlegs.

Stilt is a freshwater specialist, preferring these habitats to saltwater locations, and its long legs and neck, small pigeon-like head, and pot-bellied appearance give it a distinctive profile, especially if the observer knows what to look for. Similar to dowitchers and unlike yellowlegs, Stilt is a tactile feeder, sometimes foraging in loose, single-species flocks of 4–30 birds. However, they are also found among dowitchers and yellowlegs, where they are easily overlooked in their plain gray nonbreeding plumage.

Even an astute and experienced observer as Arthur Cleveland Bent allowed that, "Strangely enough, I have never seen, or rather recognized, a stilt sandpiper in life." The challenge is considerably reduced when observers

◁ These distinguished, long-legged sandpipers breed in marshy coastal tundra from extreme n. Alaska to w. Hudson Bay. This male with a very short bill has the most complete breeding plumage Kevin has ever seen, even after working on Alaska's North Slope as a biologist for four years where Stilt Sandpiper is **common.** CHURCHILL, MANITOBA, JUNE

▶ In migration and winter, Stilt Sandpipers occasionally feed alongside Lesser Yellowlegs (right), dowitchers, and Dunlin (front). Feeding Stilts probe with a rapid stitching motion with heads lowered and tails raised. In contrast, yellowlegs pick randomly and dowitchers probe methodically. These Stilts in transition plumage were photographed in early April in Texas.

▲ Stilt Sandpipers (*right*) in late July are transitioning out of breeding plumage. They forage by walking slowly with bills pointed at the water, and they often submerge their heads and raise their tails while probing the underwater substrate for invertebrates and larvae. Short-billed Dowitchers (*left*) usually stay in one place and work the substrate thoroughly in sewing machine fashion.
JAMAICA BAY WILDLIFE REFUGE, NYC

realize that when these species feed together, they forage differently, and this difference is evident even at great distances if you know what to look for.

Dowitchers probe with smooth, repetitive plunges like a sewing machine, and they mostly stay in the same feeding area for long periods. Stilt Sandpiper stitches the substrate with a rapid series of jabs, often with heads totally submerged, and they more actively move around while foraging. Lesser Yellowlegs, as a heads-up sight hunter, actively picks at the water or substrate, while Stilt forages by methodically walking with head and bill pointed toward the water, similar to Little Blue Heron (see photo above).

Breeding exclusively in North America in a handful of segregated Arctic and subarctic tundra locations, Stilt nests locally in open, moist sedge habitats in high Arctic locations near the Beaufort Sea in Alaska and Canada, but also in lower subarctic regions near Hudson Bay, where they nest among stunted Dwarf Birch and willows with scattered Black Spruce. They continue to be widely studied in two small breeding ranges on the west and southwest side of Hudson Bay, where access to their breeding territories is relatively easy, making them one of the most studied of Arctic shorebirds.

Stilt is a locally common long-distance migrant along the Texas coast and through the Great Plains in spring and fall, and along the Atlantic Coast in fall. It is scarce in the West. In migration and on breeding grounds, the primary prey is insects and their larvae associated with aquatic environments, including diving beetle, midge, and mosquito larvae. Since they forage mostly by touch using their sensitive bill tips, Stilts often feed at night. Vegetable matter, including seeds, is another important food source.

Spring migrants depart South American wintering areas between late February and mid-May, with arrival on the Gulf Coast in March. Small numbers pass through the se. United States mostly from late April to mid-May. While often seen in small to medium flocks, Kevin once counted more than 1,000 Stilt Sandpipers in one large pool at the Anahuac NWR in Texas in the second week of April, and most of these birds had moved north by the next week.

Birds staging on the coasts of Suriname and Venezuela in spring typically overfly the Gulf of Mexico. Most birds, however, beginning with adult males, migrate north through Texas and Louisiana in March and early April, where they spend time feeding in productive wetland habitats. From here they move north through the Great Plains, with Cheyenne Bottoms NWR a major staging area, and peak numbers occur in late April to late May. Jumps between staging areas may involve hundreds of miles.

Stilt Sandpipers reach Churchill, Manitoba, along the west side of Hudson Bay in late May and early June, with males arriving first. Courtship begins when females appear soon afterward. Pairs show a high degree of fidelity to their territory and to each other, but if arrival times of previously paired adults is widely spaced, birds may choose a new mate. Males build several scrapes in dry locations or on small hummocks in their territories, and they straddle prospective scrapes while females crawl under them and settle in for a test run.

Migrating east of the Rockies, flying Stilt Sandpipers display long-legged elegance and a white core to gray underwings. Their flight profile, with a shorter rear body and wings and a longer bill that droops near the tip, differs from that of Lesser Yellowlegs, which has a short, thin, straight bill and longer rear body. Dunlin in flight differs from Stilt by legs that don't extend past the tail. TEXAS, MARCH

Nests are typically exposed but fairly deep, thus affording cover and protection. Pairs may use the same refurbished scrape in successive years. Both sexes incubate the 3–4 eggs for about 19–21 days, and young can fly after 17–18 days. Chicks are precocial, and parental duties involve brooding, predator detection, and leading chicks to feeding areas.

Migration is more widespread in fall than in spring, though still most numerous through the prairies and at many sites along the Atlantic Coast. Adults depart the breeding grounds in late June (failed breeders), early to mid-July (females), and mid to late July (males) and make nonstop flights to stopover sites. Peak numbers occur at stopover sites through much of the continent from late July to mid-August, with arrival at wintering grounds from early August to early September. Juveniles depart breeding grounds from late July to early September and arrive through much of the continent from mid-August to mid-September, with arrival at wintering sites mostly in late September.

Female Stilt Sandpiper ultimately selects the nest site to host her eggs, and this bird with head lowered appears well pleased with her selection. Breeding Stilts are well camouflaged and sit very tightly on their nests without flushing. Kevin's fellow biologists in Alaska actually picked one off the nest with their hands to band it. Alaska, June

▶ Handsome juvenile Stilt Sandpipers have a plumage unlike adults and might be mistaken for a dowitcher or yellowlegs. Famed ornithologist Arthur Cleveland Bent allowed: "Strangely enough, I have never seen, or rather recognized, a stilt sandpiper in life," even though he probably encountered a number of them. NYC, August

▽ A Stilt Sandpiper molting into breeding plumage (left) feeds in a tidal pool in Galveston, Texas, in April. A nonbreeding Stilt in early March (right) has a drab, mousy gray plumage with a dingy white belly and dense gray markings on the upper breast. Long yellowish legs, small head, longer, thinner neck, and streamlined body help to separate these birds from Dunlin in winter.

Winter range includes s. Florida and se. Texas in the United States, ranging south into Mexico and Central America, but most individuals migrate long distances to spend the winter in much of South America. Other wintering sites include the Bahamas, West Indies, Caribbean islands, and the Antilles. At all winter locations, Stilt Sandpiper is mostly a pond-foraging species as well as frequenter of freshwater marshes and flooded fields. In migration, the birds prefer freshwater habitats but may forage in tidal pools. They are rarely seen on tidal flats.

The North American and world population is estimated at about 1.24 million birds, based on a 2008 count in the Prairie Pothole region during northward migration (Skagen et al. 2008 in Andres et al. 2012). Declines of 6% per year along the Atlantic Coast in the late 1900s were matched by increases of 7% at more inland Eastern survey sites, leading to an assumption that the population is stable (Andres et al. 2012).

▶ Curlew Sandpiper
Calidris ferruginea

BIOMETRICS 7¼–7½ inches long; wingspan: 16¾–18½ inches; weight: 35–103 grams (1.25–3.6 ounces)
STRUCTURE Similar to Dunlin, but slimmer and more attenuated, with longer legs, neck, and wings; also has a finer, more evenly decurved bill; upright stance when at rest
STATUS Despite very large numbers, listed as Near Threatened by Bird Life International

Curlew Sandpiper is a graceful, medium-sized Eurasian shorebird with long dark legs, a longish neck, and a fairly long bill with a fine, drooping tip. There are few shorebirds as striking as Curlew Sandpiper in breeding plumage, with males showing rich rufous to brick-red underparts and face with mottled black, white, and rufous upperparts. Females have a more muted, orangish plumage. In nonbreeding

A regular vagrant along the Atlantic and upper Pacific coastlines, Curlew Sandpiper breeds on the tundra of Arctic Siberia and rarely in w. Alaska. This breeding male was photographed in New Jersey in May.

plumage, Curlew has a uniform, dull gray appearance with whitish underparts and bright white underwings and vent.

Curlew Sandpiper is a rare but annual migrant along the East Coast of North America to the Gulf of Mexico and is near annual in the Heislerville Impoundments in s. New Jersey, where it joins other northbound shorebirds on mudflats and beaches. North American sightings are mostly from May to August and involve adults and 1st-summer birds. Curlew Sandpiper occasionally augments its invertebrate diet with the eggs of horseshoe crab on Delaware Bay in spring. It also occurs with some regularity in the Pacific Northwest region during migration.

Breeding across Arctic coastal Siberia, Curlew favors open tundra with a hummocky, marshy component along seacoasts and islands. It is a very rare breeding species in coastal areas of n. Alaska and the w. Aleutian Islands, with one confirmed nest record at Point Barrow. While its huge breeding range is high Arctic Eurasia, this species seems to stray to many regions of the world outside its usual haunts. Wintering mostly in sub-Saharan Africa as well as the Middle East, Southeast Asia, and Australia, Curlew is a regular but locally uncommon winter visitor to Europe (about 1,000–1,800 individuals) and rare but regular in North America, mostly in spring and summer.

A quartet of Curlew Sandpipers in Portugal in April. A breeding male is escorting three females during their long flight to Arctic breeding grounds. While somewhat similar to Stilt Sandpiper, Curlew has dark versus yellowish legs in Stilt and more complete rust color on the head and breast (males).

▲ A high breeding-plumaged male Curlew Sandpiper appears with Semipalmated Sandpipers at Heislerville, NJ, where it is mostly an annual visitor in May. The nonbreeding Curlew in flight from Thailand in January (*right*) differs from Stilt Sandpiper by feet that don't extend much beyond the tail, and by its more slender, evenly curved bill.

Curlew is similar to Dunlin in body size and bill length, but Curlew's body is slimmer, more attenuated and tubular, with longer legs and wings and a thinner, more evenly drooping bill. It also resembles Stilt Sandpiper in size and shape but shows shorter black rather than yellow-green legs.

Curlew feeds by wading and probing, but also by patrolling drier shorelines away from other feeding shorebirds. It has a unique migration pattern that involves a handful of important staging areas, where it may stay up to three weeks or more as it fattens up and replaces critical flight feathers in an asynchronous fashion, depending on energy levels. In this scenario, it may replace more critical outer primaries while retaining less important inner ones.

With a global population estimate of 1.1–1.3 million birds, it is accounted one of the most abundant wintering shorebirds on the West African coast (350,000–450,000 birds). While population numbers are relatively high, habitat destruction, illegal hunting, and pollution in various parts of its wintering range and migration routes resulted in Curlew being listed as Near Threatened (Malpas et al. 2023). Because of its huge breeding and wintering range, population trends and estimates are very difficult to quantify.

▶ Curlew Sandpiper uses several stopover sites for a number of weeks each during its protracted migration. During these long stopovers, it replaces crucial flight feathers in asynchronous fashion, unlike other shorebirds which molt their wing feathers in order starting at the innermost secondary. This bird stayed on this Stone Harbor, NJ, beach for three weeks, from late July to August, during which time it replaced its secondaries (inner wing flight feathers) and then the outer five primary feathers, the most crucial for flying long distances, skipping the inner primaries. Semipalmated Sandpiper, Sanderling, and Ruddy Turnstone are at right.

▲ Three nonbreeding Curlew Sandpipers from Thailand in January (*left*) differ from Dunlin by having longer, dark legs, a much thinner bill that decurves evenly to the tip, and cleaner white underparts. It differs from Stilt Sandpiper by having dark versus yellowish legs in Stilt. The female at right is molting into nonbreeding plumage in August in NJ, with a Western Sandpiper nearby.

▶ Sanderling
Calidris alba

BIOMETRICS   Size: 7½–8 inches long; wingspan: 16–18 inches; weight: 40–100 grams (1.4–3.5 ounces)
STRUCTURE   Chunky body shape with large head, thick neck and stout medium-length bill; lacks rear toe
STATUS   Common along sandy coastlines; population data suggests an apparent decline, with Delaware Bay counts indicating a long-term decline.

Say the word "sandpiper" and this small, frenetic bird of sandy beaches is what springs to mind. Its likeness is emblazoned on half the gift cards, lampshades, and kitchen towels sold at seaside resorts, and in hotel room artwork. It is probably the most well-known shorebird in the world, even to nonbirders, because of its high visibility as it chases waves on most beaches around the world, including heavily inhabited areas.

Fist-sized and fist-shaped and overall pallid with a particularly whitish face (nonbreeding plumage September–April), these larger than "peep" sandpipers are found mostly in single species flocks of 5–100+ birds foraging as a unit and chasing the receding wave wash, then racing ahead of the next oncoming surge. The impression is one of birds playing tag with the waves, and more than half of a Sanderling's daily time budget is spent foraging in the tidal zone. This is one of the more consistent flocking shorebird species, and they feed, fly, roost, and escape from danger in tight knit groups that may number hundreds of birds.

◀ A very well-nourished male Sanderling which has fattened up on horseshoe crab eggs in NJ in May strikes an almost regal pose with its rust and black coat of breeding feathers. Note the lack of a rear toe on this bird, which all other small to medium sandpipers have.

Two breeding Sanderlings fly at left with strong white wing stripes and stocky bodies in May, NJ. This species is typically found in small to very large flocks, especially during migration (*right*, September) and winter, when they show a clean white and gray overall plumage.

If you look closely, you will see that Sanderlings are actually picking up stranded invertebrates in the sand as waves retreat, with their hurried escape from incoming waves a survival mechanism. Western Sandpipers occasionally join feeding Sanderlings, but the Sanderling's larger size is manifest. To facilitate running at a Sanderling's typical speed, the bird lacks a hind toe, unlike all other small to medium-sized sandpipers.

Sanderling is a long-distance migrant, and different populations travel from 3,000 to 9,000 miles between breeding and wintering grounds. Wintering almost exclusively on coastal sandy beaches with wave action, Sanderlings may stop over on alkali lakeshores and muddy or pebbly shorelines of large lakes in interior regions during migration. However, sand is Sanderling's substrate of choice, giving rise to the bird's common name, taken from the old English "sand-yoling," or "sand-ploughman." In some southern coastal areas, Sanderlings feed in nearby inland impoundments and tidal pool shorelines.

Spending all but late May, June, and early to mid-July on winter territory or in migration, Sanderling might very arguably be considered a coastal resident, being absent only as long as it takes to breed in the high Arctic reaches of the Canadian Archipelago. During a 2½ month period, they may fledge 1–3 broods (with males raising the first clutch and females the second). If a third clutch occurs, a different male raises that one. After all this busy work, they return to the same favored stretch of beach along the Atlantic, Pacific, or Gulf shorelines.

Many North America breeders spend the winter along the coasts of South America south to Tierra del Fuego (50°S), with others ranging as far north as Massachusetts and southern British Columbia (50°N). Sanderlings enjoy one of the greatest latitudinal winter ranges of any bird species, spanning 100 degrees of latitude and encompassing temperate and tropical regions. A bird's winter territory is more modest and includes the wave zone along a few miles of ocean beachfront and adjacent tidal locations.

Sanderling is everyone's image of a shorebird—small birds that play tag with the waves. If you watch carefully, they are picking up stranded invertebrates while chasing the wave back to the sea but then running for safety as waves crash in again. NJ, DECEMBER

Sandpipers and Allies (Family Scolopacidae)

Sanderling's stout bill is ideal for picking and light probing with stitching jabs as it pries small bivalves, crustaceans, and polychaetes from the wave-churned slurry before the sand compacts or the next wave arrives, sending the birds racing to higher ground. While most sandpipers probe for food, there is evidence that Sanderling also use visual clues to locate prey.

One of its favorite wave zone foods is mole crab, a soft-shell crustacean that Sanderling locates by the bubbles emitted from the crab's burrows as the waves retreat. When one of these relatively large, high-protein food items is secured, multiple Sanderlings aggressively chase the successful hunter, trying to steal its treasured meal. Roosting birds gather in tightly packed flocks of 40 or more on sand or cobble beaches above the reach of waves. During migration at a handful of select stopover locations, roosts may contain hundreds, even thousands of individuals.

Spring migration takes place between early March and mid-June. Sanderling is one of three species, along with Red Knot and Ruddy Turnstone, that make up a large percentage of shorebirds that feed on horseshoe crab eggs for critical sustenance during spring migration along the Delaware Bayshore in May. Tens of thousands of Sanderlings rely on this important source of protein to allow them to successfully navigate the several-thousand-mile journey to high Arctic breeding areas.

Birds arrive on Arctic breeding grounds from late May to mid-June, with snow typically still covering the open,

A typical flock of Sanderlings just starting to molt into breeding plumage on a St. Augustine, Florida beach in early April. Can you spot two other shorebirds in this photo? A breeding Black-bellied Plover is at rear left, while a smaller Semipalmated Plover is hiding in the pack at front right. Note the head of a Least Tern at rear center; as the smallest tern in the world, it is the same size as a Sanderling, but with longer wings.

Sanderlings in an array of plumages join Red Knots and a few Ruddy Turnstones in a feeding frenzy on New Jersey's Delaware Bayshore in May. Breeding males show a rich, rust plumage while females are typically more silvery. First-year males show a plumage in between the two. Can you spot the lone Semipalmated Sandpiper in the crowd? Hint: look left of center. And the lone Dunlin; another hint: left of the Ruddy Turnstones.

▲ Unlike many sandpipers, Sanderlings truly prefer to forage on sandy substrates. Most "sand" pipers are really "mudpipers." A breeding male in late April in Florida (*left*) shows typical rich, rust and black plumage while nonbreeding birds represent our only true "white" sandpiper in winter, though touches of pearl gray adorn its back. TEXAS, SEPTEMBER

stony, sparsely vegetated tundra that Sanderlings favor. Their shallow cup-shaped nest is constructed by the female out of dry leaves, bits of moss, and lichens. Both sexes arrive on the breeding grounds simultaneously, and within a short time are paired. Both adults share parental duties, and incubation takes 23–27 days. Young can typically fly after 17 days.

At nesting areas, both sexes are garbed in breeding plumage, with males showing a striking brick-red head, neck, and breast with touches of red on the back, and females having a more muted plumage that can be either rusty or mostly silver-gray. South of breeding areas, this reddish breeding plumage is seen only in May/mid-June or late July/early August, so many observers are familiar only with Sanderling's pallid, whitish-gray nonbreeding plumage, the foundation of its scientific name, *C. alba* (Latin for white). In all plumages, Sanderling shows a bold

white wing stripe against mostly blackish upperwings in flight, and their call is a sharp, metallic *plink*. Despite its widespread global distribution, no subspecies exist.

Fall migration takes place between late June and mid-November. Southbound adults depart breeding areas mostly between mid-July and mid-August, but failed breeders may begin to head south by late June. Molt from breeding to nonbreeding plumage occurs from July to September, with a salt-and-pepper look to the upperparts and a blush of orange on the face and upper breast. Most juveniles depart the tundra by early to mid-August and gather in flocks at nearby coastal areas. They begin to head south by late August and early September.

While Sanderling is one of the most commonly seen shorebirds, its worldwide population is apparently declining. The global population is estimated at 700,000 birds, with about 300,000 in North America (Morrison et al.

▼ Juvenile Sanderling's plumage (*left*, August, NJ) is one of the neatest and crispest of all shorebirds, with a smooth contrast of various shades of black and white. A molting adult at right in late July is shown with a smaller, worn Semipalmated Sandpiper whose feathers have seen better days and need replacing.

▶ This high contrast photo shows similar-sized Sanderling and Dunlin flying over a mirror surf line in February in NJ. Note the strong white wing stripe on both birds, but that is where the similarities end.

2006). While shorebird numbers are often difficult to determine because the birds cluster in large groups scattered over great distances, a few surveys indicate sharp declines in Sanderling numbers at some locations, while recent counts in Peru suggest that populations may be higher than previous estimates (Andres et al. 2012).

During 1974–1982, Sanderlings on the Atlantic Coast dropped by 13.7% per year, and peak migration counts from Massachusetts during that same period dropped from 10,000–30,000 in the 1950s to average around 2,000 birds. Counts from migration surveys and Christmas Bird Counts are highly variable among years and show no increasing or decreasing trends since the 1970s (Butcher and Niven 2007 in Andres et al. 2012). Information along Delaware Bay indicates a long-term decline (Niles et al. 2009 in Andres et al. 2012), so a precautionary similar decline is suggested (Andres et al. 2012).

▶ Dunlin
Calidris alpina

BIOMETRICS 6½–8¾ inches long; wingspan: 12¾–17½ inches; weight: 48–64 grams (1.7–2.2 ounces)
STRUCTURE Chunky with short, thick neck, short black legs, and longish bill with a thick base that droops near the fine tip
STATUS Most abundant wintering shorebird in many coastal locations; scarce migrant at mostly inland sites; *C. a. pacifica* and *C. a. hudsonia* populations are considered stable in short and long term; *C. a. arcticola* declined about 30% from 2006 to 2012 (Andres et al. 2012).

- -

As the most widespread and one of the most numerous of all North American shorebirds, especially in coastal areas, Dunlin is very familiar among birders who take the

◀ Once named Red-backed Sandpiper, Dunlin is one of the most abundant and widespread shorebirds in the Northern Hemisphere. In North America, they breed in Arctic and subarctic portions of Canada and Alaska. This stunning male belongs to the Eastern subspecies *C .a. hudsonia.* NJ, MAY

Resting Dunlin with Semipalmated Sandpipers near Stone Harbor, NJ in May. Preferring to feed on expansive tidal mudflats, these birds are sitting out the high tide in a quiet, trapped tidal pool along the Atlantic coastline.

time to study shorebirds. Formerly called "Red-backed Sandpiper" as recently as the mid-20th century because of its striking breeding plumage with a vivid rusty back and black belly patch, it is now named for its nonbreeding plumage, a mousy gray/brown or "dun" color.

Dunlin typically gathers in flocks numbering from a few dozen to a few thousand, when they are easily viewed at close range along seacoasts and tidal flats, or in flooded agricultural fields at inland locations during migration. Walking deliberately and ranging over a wide area while foraging, Dunlin feed with an up-and-down sewing-machine motion by inserting their long bill with a drooping tip into the substrate, typically in shallow water less than a few inches deep.

So ripe with superlatives is this medium-sized sandpiper that it is difficult to choose which special trait to highlight first. So, starting broadly, Dunlin is a cosmopolitan species that breeds in Arctic and subarctic tundra habitats in Europe, Asia, North America, Greenland, and Iceland and winters coastally north of the equator in Africa, southeast Asia, and North America.

No other shorebird is more polytypic (having many taxonomic subdivisions), with up to 9 subspecies recognized worldwide, although the taxonomic status of several named subspecies is open to discussion. The American Ornithological Society (AOS) recognizes a single North American subspecies, though it recognizes three distinct and segregated breeding populations. Many authorities recognize three subspecies in North America, with Nils Warnock and Frank Gill (1996), two of North America's most esteemed shorebird biologists, fully supporting this idea, as do Kevin Karlson and *The Shorebird Guide* (O'Brien et al. 2006).

A breeding male Dunlin of the subspecies *C. a. articola* stands guard near his high Arctic tundra nest on the North Slope of Alaska in June. Birds from this area winter mostly in e. China, Korea, and Japan after breeding, and it is the only subspecies in North America to have declined by about 30% from 2006 to 2012.

▲ Resting Dunlin of the subspecies *C.a. pacifica* present a peaceful scene during migration in early May, but this tranquility is often interrupted by predators such as Merlin and Peregrine Falcon, so some birds keep an eye open for danger, even when catching a few winks. CORDOVA, ALASKA

▽ In Point Barrow, Alaska, a territorial Dunlin is using a snow mound to express its angst over intruders or another male Dunlin in its breeding territory in June. Note the clean white underwings and a defined black belly-patch.

The three subspecies are:

- *C. a. arcticola:* breeds in nw. Alaska and nw. Canada and winters mostly in e. China, Korea, and Japan.
- *C. a. pacifica*: breeds in sw. Alaska and winters in Pacific coastal areas from Alaska to central Mexico.
- *C. a. hudsonia*: breeds in north-central Canada and winters along the Atlantic and Gulf coasts from New England to ne. Mexico.

Dunlin is a short- to medium-distance migrant, and they are among our earliest spring migrants. *C. a. arcticola* and *C. a. hudsonia* begin to migrate in mid-March, while *C. a. pacifica* begins to depart wintering areas in January. These three subspecies have different migration timing periods with respect to arrival and departure dates, but nest initiation is primarily early June (*pacifica*) and mid-June (*arcticola* and *hudsonia*), and some birds may arrive at more southern subarctic breeding areas in early May. In years marked by late snowmelt, they may arrive already paired.

Dunlins are monogamous, with both parents attending their single brood through the nesting season, although females typically desert the male and young after roughly 6 days. The 3–4 eggs are often concealed in a tussock in

△ Well-camouflaged eggs in a Dunlin nest are shown at left in Kuparuk, Alaska, in June. Dunlin nests are well hidden by surrounding grasses, and incubating birds are very hard to see, as shown in this typical nest.

moist sedge tundra with standing water nearby. Incubation is 21–22 days, and young can fly 18–24 days after hatching. Outside the breeding season, Dunlin concentrates in tidal estuaries with exposed mudflats, on the edges of impoundments and other water bodies, and on sandy beaches at high tide.

Few shorebirds demonstrate so marked a transition between breeding and nonbreeding plumages, with both males and females donning rich red backs and jet-black bellies when breeding and toning down to the blandest of mousy brown upperparts with dull white bellies and gray throats before migrating south in vast numbers. Males have bolder, more complete breeding plumage and shorter bills.

These hardy birds are among the last shorebirds to migrate in the fall, and adults and juveniles often migrate together. Fall migration varies among subspecies. *C. a. arcticola* departs breeding grounds and moves to nearby coastal staging areas from early July to early September. Most birds then move to coastal w. Alaska in September/ October, where they mix with *pacifica* birds before departing to se. Asia wintering areas. Adult *pacifica* Dunlins depart breeding areas and concentrate in nearby coastal areas from late June to early November. Most of these birds depart abruptly in early October and make single transoceanic flights to coastal areas from s. British Columbia to central California, with some birds drifting south and inland to wintering areas in Mexico and California's Central Valley.

▷ Dunlin form small to very large flocks numbering in the thousands in winter. They excel at close-quarter aerial maneuvers, and you can almost hear the sizzle of their collective wings.
NJ, FEBRUARY

Adult *hudsonia* birds depart breeding grounds to staging areas along Hudson and James Bays between July and early September, with peak numbers passing through New England and the mid-Atlantic in October and November.

Dunlin winter farther north than most shorebirds and penetrate no farther south than the midway point of both Mexican coasts. In winter, Dunlins swarm over coastal mudflats like a hungry tide, ever probing for benthic prey but ever mindful of hunting hawks and falcons, which may trim down flocks over the course of the winter. Dunlins' response to aerial predators is to gather in tight, wheeling flocks that shift and climb like a roller coaster as they try to escape the predator. Able to reach speeds in excess of 100 mph, they make the air sizzle when flocks pass by with their conjoined wingbeats.

While certainly one of our most numerous shorebirds, with the North American population estimated at 2.15 million birds, the Pacific population (*C. a. arcticola*) has declined about 30% since 2006 and now numbers around 500,000. The Pacific Coast wintering Dunlin, mostly *C. a. pacifica*, and the eastern *C. a. hudsonia* population are considered stable in the short and long terms, and both these populations make up about 1.65 million of the 2.15 million birds in North America (Andres et al. 2012).

As one of the last Arctic shorebirds to migrate south, Dunlin in almost complete juvenile plumage like this bird in September, NJ (*left*) are infrequently seen in the continental United States. First-winter birds (*right*) show variable plumages into midwinter, with differing amounts of black on the underparts. NJ, OCTOBER

After bathing, a breeding-plumaged Dunlin in NJ in May does a lift-up to shake water off its feathers (*left*). A 1st-winter bird in flight in November shows a few retained rusty juvenile feathers on its scapulars (upper back) and a distinct white wing stripe.

▶ Rock Sandpiper
Calidris ptilocenemis

BIOMETRICS 7¼–9½ inches long; wingspan: 13½–18½ inches; weight: 71–114 grams (2.5–4 ounces)
STRUCTURE Chunky and rounded with a short, thick neck, short legs, and medium to longish drooped bill; horizontal crouched posture
STATUS Fairly common but highly localized along seacoasts; less common in southern part of winter range; population trends for all subspecies is unknown

----

As one of the hardiest of all shorebirds, Rock Sandpiper evolved as the Beringian Land Bridge that supported it was de-evolving, leaving isolated populations of the subspecies *ptilocenemis* on geographically separated islands on both sides of the Bering Sea. This segregation of the species resulted in one of the most variable of all sandpipers with respect to size and color patterns, enough to warrant four or five subspecies. This distinction makes Rock Sandpiper one of the most polytypic of all the small-to medium-sized sandpipers.

Three subspecies breed in North America, all on Alaska's west coast and offshore islands:

■ *C. p. ptilocenemis*: breeds on Hall and St. Matthew's islands as well as the Pribilof Islands and winters on the Alaskan Peninsula
■ *C. p. couesi*: breeds and winters on the Alaskan Peninsula and the Aleutians, where it moves to the seacoasts in winter

■ *C. p. tschuktschorum*: breeds in w. Alaska, St. Lawrence and Nunivak islands, and the Chukchi Peninsula (Russia), and winters in Cook Inlet and se. Alaska south to n. California. This northernmost subspecies winters the farthest south of all Rock Sandpipers.

The larger, paler race (*C. p. ptilocenemis*) that breeds on the Pribilof islands was once known as the Pribilof Sandpiper, *Arquatella ptilocenemis ptilocenemis*, and the smaller, darker Aleutian race, *C. p. couesi*, was known as the Aleutian Sandpiper (Bent 1962). It wasn't until 1883 that Rock Sandpiper was separated from Purple Sandpiper, whose more eastern breeding range likewise comprises disconnected breeding areas.

This sturdy, plumpish *Calidris* sandpiper winters farther north than any other North American shorebird except Purple Sandpiper, with most populations relocating to better winter-feeding areas within their region. The one exception is the Aleutian race, *C. p. couesi*, which is a year-round resident of that area. Only *C. p. tschuktschorum* is fully migratory, with Alaskan and Siberian birds relocating to rocky coastlines from se. Alaska to n. California, a migration that for some individuals may exceed 5,000 miles. Each of the subspecies has a distinctive breeding plumage, but a similar nonbreeding plumage that is comparable to Purple Sandpiper, *C. maritima*.

Rock Sandpipers that winter in the Pacific Northwest to n. California spend much of their time foraging in close association with Black Oystercatchers, Surfbirds, Black Turnstones, and Sanderlings. The different body sizes and

▶ A breeding Rock Sandpiper of the subspecies *C. p. ptilocenemis* is shown on the Pribilof Islands in June. This race winters mainly on the Alaskan peninsula, showing a hardiness that is unique among shorebirds. Rock Sandpipers winter farther north than any North American shorebirds except Purple Sandpiper.

▶ Where winter ranges overlap, Rock Sandpipers often team up with Surfbird (*left*) and Black Turnstone (*right*), "rockpipers" all. This photo from Half Moon Bay in California in November is close to the southernmost Pacific Coast winter range for this species.

▼ A single-wing wave is part of courtship displays for most Arctic shorebirds, including this male Rock Sandpiper from the Pribilof Islands, Alaska, in June. Note the clean white underwings, and you can almost hear the monotone, droning sound coming from the bird's throat.

Rock Sandpiper breeds on low-elevation heath tundra in coastal areas, often along streambeds and wet meadows. Nest initiation is primarily mid-May in the south and early to mid-June in the north. Their energetic and vocal courtship makes them one of the most conspicuous birds on the tundra, with most birds arriving in small flocks from mid to late May, and with some birds still in nonbreeding plumage. They show a great deal of fidelity to nest sites used the previous year, where newly arrived males establish territories in snow-free areas that expand as snow melts. Flight displays both define territories and attract females, who examine the array of scrapes constructed by the male.

Nests are often in concealing cover, and nest building is an important component of pair bonding. Cups are deep with a thick layer of willow leaves, sedges, and lichen at the bottom. Both parents incubate eggs, and the pair bond continues until the female departs before the brood fledges. Incubation is 20–24 days, and young can fly 20–22 days after hatching. By late August, adults and juveniles join flocks of other shorebirds along outer coasts. From this point onward for most Rock Sandpipers, theirs is a life wedded to rocky coastlines and kelp-covered shelves, where they forage within the splash zone for an array of invertebrate prey, especially barnacles, mussels, and snails.

Despite Rock Sandpiper's common name, some contrary populations spend the entire winter on soft substrates devoid of rocks, but rich in benthic organisms, especially clams. Most birds make only modest relocations for winter, moving from inland breeding areas to coastal locations. In prime staging areas of tidal estuaries on the Alaskan Peninsula or Yukon-Kuskokwim Delta, tens of thousands of Rock Sandpipers may gather to molt in mid to

bill shapes of these birds allow them to concentrate on different kinds of prey in the same area without competing with each other. In spring, *C. p. tschuktschorum* departs wintering areas in California between mid-March and mid-May, with peak numbers passing through British Columbia from mid-April to early May. Arrival on breeding grounds takes place mostly in mid to late May.

▲ Both this molting adult (*left*) and juvenile (*right*) belong to the highly migratory subspecies of Rock Sandpiper *C. p. tschuktschorum* and were photographed in Alaska in late August. GAMBELL (*LEFT*); NOME (*RIGHT*)

late July, remaining until early October. As Arctic breeders go, Rock Sandpiper ranks as the earliest spring (April) and latest fall migrants (October–November) of all.

Rock Sandpipers that winter in northern waters (most notably *C. p. ptilocenemis*, which spends much of the winter in Cook Inlet near Anchorage, Alaska) face challenges relating to food acquisition when plummeting air temperatures cause the tidal mudflats to freeze. This puts the bird's primary winter food resource, the tiny but incredibly abundant macoma clam, out of the reach of their bills. These resourceful birds then forage at the interface between the water and the mud, where scouring ice dragged over the mud by a falling tide gives birds precious seconds to forage. During severe winter cold spells, birds regularly move as many as 100 miles to find feeding areas and return when temperatures rise.

Rock Sandpipers also gather in massed flocks that shelter individuals in the flock's interior from icy, heat-leaching winds. The spherical shape and larger body mass of *C. p. ptilocenemis* also help conserve body heat (Bergmann's rule), and birds increase their mass in fall by 40%, mostly in fat that helps insulate and sustain them during extremely

▼ This unique photo of nonbreeding Rock Sandpipers in Alaska in February by Brian Guzzetti shows the palest subspecies *C. p. ptilocenemis* at right, and two large, dark birds at left that are likely *C. p. tschuktschorum*. The smallest, dark bird in center is possibly the resident race *C. p. couesi*, but variation precludes certain subspecies identification.

cold or stormy stretches. Shorebirds generate heat by shivering, and enlarged pectoral muscles assist birds in this regard in winter.

There is also evidence that Rock Sandpiper's liver becomes enlarged in winter, thus speeding the absorption of nutrients. The contour feathers of *C. p. ptilocenemis* have a downier base than comparable sandpipers, especially in winter, and unlike those of most sandpipers, the feathers on Rock Sandpiper cover most of the tibia, or upper leg. In short, this is a sandpiper designed for cold conditions that other shorebirds migrate thousands of miles to avoid. A hardy bird indeed!

The total population estimate for the three subspecies that breed in North America is around 145,000 (*C. p. tschuktschorum*, 50,000; *C. p. couesi*, 75,000; and *C. p. ptilocenemis*, 19,832) (Ruthrauff et al. 2012 in Andres et al. 2012). Trends in all populations are considered unknown. The population of *C. p. kurilensis* is believed to number a mere 200–300 pairs and is designated Threatened in Russia.

▶ Purple Sandpiper
Calidris maritima

BIOMETRICS 8–8¾ inches long; wingspan: 16¾–18½ inches; weight: 52–106 grams (1.8–3.75 ounces)
STRUCTURE Chunky and round bodied with short, thick neck and short legs; medium to longish drooped bill; low, crouching horizontal posture
STATUS Fairly common but local along seacoasts; rare but regular visitor in migration to Great Lakes and Gulf Coast; population probably stable, but hard to determine because of remoteness of wintering areas of this species

- -

You are alone at the end of a rock jetty, casting metal into a choppy, lead-colored sea. Suddenly you find yourself surrounded by a squadron of lead-colored birds scampering about the rocks. They seemed to appear out of nowhere, but reason dictates they were only resting on an adjacent jetty and then flew in so low they were invisible against the winter sea. Crowding close to the water's edge, they seem

◁ Purple Sandpipers, like this adult in NJ in January, are slate gray above with a gray hood and gray streaks and spots below. While this adult bird shows no purple highlights on its upperparts, a number of nonbreeding birds do (see page 181, bottom photo).

▽ Purple Sandpipers form small- to moderate-sized flocks in winter, when they frequent rocky shorelines and jetties. Their range includes extreme s. Greenland, where they are resident, and from Newfoundland to n. South Carolina. Can you find the one Dunlin in this flock in NJ in February? Hint: it is smaller and paler and at the edge of the flock.

While Purple Sandpipers are typically seen in tight flocks while feeding on surf-washed rocky habitats in winter, they occasionally join Ruddy Turnstones and Dunlin. This photo was taken on a long, rocky jetty in Wildwood Crest, NJ in December.

heedless of the onrushing surf that could dash them against the rocks—that is, until a larger than average wave rolls in, and the birds rise like sea spray to coalesce offshore before returning to the jetty from which they were displaced.

Nah, nah, nah, nah, nah, your waves will never catch me, is what they seem to be saying with focused indifference. Indeed, it is very unusual to see one of these nimble birds tagged by a wave. Flushed, yes, but engulfed, no. These scampering "rockpipers" are masters of their risky trade, scouring the high-energy wave zone of rocky shores for prey beyond the reach of other Eastern coastal wintering shorebirds.

Ruddy Turnstone, Dunlin, Sanderling, Red Knot, and Black-bellied Plover often sit out the high tide above the splash zone with Purple Sandpiper, but when the tide falls, only *C. maritima* (the sandpiper that belongs to the sea) is able to exploit the invertebrate riches within the wave zone. Occasionally Ruddy Turnstone and Dunlin stray into this perilous wave zone to feed near the ocean, but a few bad washouts usually discourage this behavior.

As for hardiness, Purple Sandpipers range as far north as coastal Greenland and Iceland in winter. Wintering farther north than any other shorebird, Purple Sandpipers that breed in southern Greenland are believed to be resident, and in winter, their ranks are joined by birds that breed in northern Greenland. Most people know this chunky intrepid "rockpiper" from its slate gray nonbreeding plumage, often with rufous fringes to the scapulars.

To see the birds in their rufous-tinged breeding plumage, you could make a journey to their breeding grounds in coastal Greenland, on the shores of Hudson Bay, or in the northern reaches of Nunavut Province in Canada, or you could visit the coastlines of Labrador and Newfoundland in May. You can also see mostly full breeding-plumaged birds on rocky jetties and substrates along the upper US Atlantic coast from New Jersey northward in early May, but this usually involves 1st-year birds with retained worn juvenile wing coverts from the previous year. This high Arctic and subarctic nester also breeds in Iceland, northern Europe, and Russia.

Yes! This nonbreeding Purple Sandpiper from NJ in February shows a good amount of the purplish color on its upperparts that is the namesake of this species. Short, sturdy legs and a low center of gravity help Purple Sandpipers navigate the slippery, surf-washed rocks that are their winter home.

◁ Purple Sandpiper's breeding plumage is rarely seen in southern wintering areas, but a number of 1st-year birds that molt into almost complete breeding condition occur along the upper Atlantic Coast in early May. This mostly breeding-plumaged 1st-year bird from NJ in early May lacks the purple edges to upperpart feathers seen in adults, and it shows retained, worn juvenile greater wing coverts, tertials, and primaries that help to age it.

Not particularly early migrants, birds depart wintering areas on the Atlantic Coast in April–May and reach breeding areas in the first half of June. Males establish territory within days of arrival by utilizing aerial displays. Adults return to the same breeding areas each year, and paired birds appear to mate for life, an advantage where breeding seasons are short and established pairs can get down to business of procreation quickly.

Once territories are sufficiently free of snow and ice, pairs begin nest building, with some birds in poorly drained areas forced to wait until early July to breed. Males make initial scrapes in high alpine tundra with moors or extensive ground cover, with sloping or flat terrain that is likely to remain moist through the summer. After selecting a scrape, females make improvements by adding a layer of dried willow leaves and perhaps a few down feathers. Primarily the male incubates the 4 eggs for 21–22 days, and he alone tends to the chicks, which can fly after 3–4 weeks.

Eggs are amazingly hardy and can be left unattended for up to 24 hours in subfreezing temperatures with little or no ill effects. Chicks are mobile within hours of hatching and able to forage from Day 1. Males continue to brood young for 7–9 days, and they attend juveniles for 28–34 days, at which time the parent-brood bond ends. Females and males whose nests fail begin migrating to far northern staging areas by mid-July, with other successful males following in August. Most birds depart by mid-September, and wintering areas are typically reached from October to December.

Breeding Purple Sandpipers that do not winter in Europe via Iceland occur on rocky coasts from Newfoundland south to North Carolina from late fall to midspring. South of New England's natural rocky coastlines, Purple Sandpipers gather in flocks of 20–40 birds on man-made rock jetties, breakwaters, and sea walls along sandy coasts, often roosting with Sanderling and Dunlin.

In winter and during migration, some birds also occur in proper habitat at interior locations, including along the

◁ Seeking respite from the rising tide and crashing waves, Purple Sandpipers (with darker bodies) are sitting out the tide with Sanderlings and Dunlin, two other hardy winter coastal residents. If you look carefully, you will spot three Ruddy Turnstones in this pack from NJ in early April. Purple Sandpipers winter farther north than any other North American shorebird (s. Greenland).

△ Food opportunities like this beached Quahog clam lured this wintering Purple Sandpiper away from its typical rock jetty habitat and onto the sand in Cape May, NJ in March. Almost always found in flocks, the birds flying at right in NJ in December are moving between jetties.

shores of the Great Lakes. The primary winter fare of Purple Sandpiper includes snails and bivalves (especially mussels and periwinkles), as well as crustaceans, insects, marine worms, and some algae. Purple Sandpiper is a short to medium-distance migrant, with more northerly breeders apparently migrating southeast to winter in Greenland and Europe, while southerly breeders migrate across Quebec and Labrador to winter on the American side.

On the granite coast of Maine, an important wintering area, they occupy offshore islands and peninsulas and are frequently seen in the intertidal zones of rocky ledges feeding amid seaweed. Only rarely is this species encountered on anything but rocky substrate, though Kevin once saw a number of Purple Sandpipers feeding on a sandy beach in Cape May, NJ following a stranding of large northern quahog clams, and the birds were enjoying an opportunistic, hearty meal.

Two subspecies occur in North America: *C. m. belcheri*, which breeds on the Belcher Islands and around Hudson Bay; and *C. m. maritima*, which breeds elsewhere in the nc. Canadian Arctic to Iceland and Greenland. Global population estimates for this species range from 205,000 to 295,000 birds (Malpas et al. 2023: Purple Sandpiper), with about 25,000 birds from the two subspecies breeding in North America. Christmas Bird Count numbers suggest a long-term decline of 1.8% (Butcher and Niven 2007 in Andres et al. 2012), though a substantial portion of the population winters on nearshore islands outside of CBC circles. Recent counts in Maine estimate a wintering population of 15,000 birds (Andres et al. 2012), and some individuals of *C. m. maritima* overwinter in Europe, making population trends difficult to ascertain.

▶ Baird's Sandpiper
Calidris bairdii

BIOMETRICS 5¾–7¼ inches long; wingspan: 16–18½ inches; weight: 28–58 grams (1–2 ounces)
STRUCTURE Slender, tubular body with long, tapered wings that give Baird's an elegant appearance; proportionally small-headed with slender, fine-tipped bill; upright stance with longish neck when alarmed
STATUS Uncommon to fairly common migrant through Great Plains and Mountain West, where it is sometimes the most numerous shorebird; scarce elsewhere in fall; population trends unknown

This dapper, crisply marked, high Arctic breeding sandpiper has one of the longest and swiftest migrations of any shorebird, as well as a huge breeding range entirely above the Arctic Circle extending more than 17,000 square miles from eastern Siberia to northwestern Greenland. Leaving high Arctic breeding grounds from late June (failed breeders) to early August (females before males), birds head directly to staging areas in the n. Great Plains and Rocky Mountains before flying on to northern South America, or directly to Tierra del Fuego, a total distance of about 9,300 miles. The amazing part of this story, however, is that they complete this remarkable journey in as little as 5 weeks.

But wait, there's more. On the return trip, females arrive in the Arctic between mid-May and early June with little to no fat reserves. After feeding nonstop for a short time, a female lays 4 eggs in just 4 days that in aggregate weigh 80–120% of her body mass. How she accomplishes this remarkable energetic feat in the seasonally food-impoverished high Arctic is tantamount to magic.

◀ The dapper, elegant Baird's Sandpiper is a member of the "grasspiper guild," shorebirds partial to foraging in moist or dry, short-grass habitats. The tasteful black, gray, and white plumage of breeding adults like this bird from the high Arctic in Prudhoe Bay, Alaska, in June would be fitting attire at a Boston Brahmin social gathering.

▼ Long wings that extend past the tail and a long, tapered rear body are trademarks of Baird's Sandpiper. The bird at left is molting into breeding plumage in Texas in April, while the one at right is molting out of breeding plumage in Arizona, early August. A smaller, fresh juvenile female Western Sandpiper is behind Baird's, and a Wilson's Phalarope is spinning in the background.

◀ A Baird's Sandpiper on the tundra in Point Barrow, Alaska, in June is performing a single-wing courtship and/or territorial display to attract a female, or to warn other Baird's that he is ready and able to defend his territory.

With long wings that evolved to accommodate very long migrations, this largest of the "peeps" has an elegant profile that makes it endearing to most birders. Their delicate buff and brown tones give them a warmer appearance than the gray and brown plumage of many other sandpipers, and one that is readily recognizable in the field. Pete poignantly describes their appearance in his fine book, *Pete Dunne's Essential Field Guide Companion* (2006), as one whose stature is akin to one who "wears a smoking jacket and smokes a pipe," a reference to the distinguished appearance of Baird's. Among the last shorebirds described to science, it was named by Elliott Coues in 1861 to honor Spencer Fullerton Baird, the second Secretary of the Smithsonian Institute in Washington, DC.

△ A Baird's Sandpiper carefully nestles onto its 4 eggs in moist, grassy Arctic tundra in Kuparuk, Alaska, in June. Baird's Sandpiper eggs rank among the planet's most richly patterned and attractive of all. Note the buckwheat leaves used by Baird's to adorn the nest lining.

Protracted spring migration begins in early March in South America, and birds overfly Central America to arrive on the Gulf Coast from mid-March to late April. From here they continue through the Mississippi River Valley and Plains states (April–May), after which they fly on to Canada, passing west of Hudson Bay, and arrive on territory from mid-May to early June. Favored breeding locations include low mountaintops, barren gravel ridges, moist grassy coastal tundra habitats, and raised beaches. Not coincidentally, these locations are likely the first to become snow free due to wind action.

Males establish territories several days after arrival by hover-circling while making trilled vocalizations, similar to other peep. Courtship and pairing begin immediately upon the female's arrival, and in some locations, birds may arrive already paired, though the pair bond is not known to persist after the breeding season. The somewhat exposed nest is constructed by both adults and consists of a shallow scrape lined with assorted dry plant material, leaves, and lichen. Some nests are unlined.

The 4 especially attractive eggs are deposited from mid-June to early July and incubated more by the male during the early nesting process. Attractive chicks hatch in 21 days, and both adults attend early broods, but the male deserts females and young after 5–7 days. Young leave the nest soon after hatching and are led to well-drained tundra and muddy lakeshores, fledging in about 20 days. Given the brevity of the breeding season, failed

▽ Though difficult to see, which is the point, a Baird's Sandpiper at left is brooding three chicks by taking them under its wings and belly. Baird's chicks are some of the most attractive of all with a pleasing mix of rust, black, and white. This little puffball is only 2 days out of the egg. ALASKA, EARLY JULY

Baird's Sandpipers lack none of the elegance of adults and may actually appear more attractive to some birders. Pale-edged back feathers impart a sense of scaliness. The left bird with very long bill and heavier body is surely a female (NJ, Sept). The right bird shows the iconic aristocratic stance that conveys a measure of class to Baird's (NY, Aug).

Adults in mostly nonbreeding plumage aggressively vie for prime foraging space, even though there is more than enough for all at Wilcox, AZ in August. Note the clean, buff appearance that differs strongly from its breeding plumage, and very long, tapered wings that allow this species to vault hemispheres and complete one of the longest migrations of all shorebirds.

efforts do not typically lead to a re-nest, although it may occur with some early failures.

Most adults depart breeding grounds from late June (failed breeders) to early August (females before males) and migrate south through the Great Plains and Rocky Mountains, with peak numbers in August. Juveniles depart breeding grounds from late July to mid-August, typically before the first permanent snowfall in far northern areas. Arrival at wintering sites occurs mostly from late September to early October.

Siberian breeders fly first to Alaska before turning south through the Great Plains, and then on to South America, bypassing Central America by flying off the Pacific Coast to the Andes. Juveniles engage in a broad-front migration

that spans the continent and includes both coasts, although a majority migrate through the interior United States. Indeed, it was juvenile plumage that Roger Tory Peterson described and illustrated in the second edition of his famous field guide (1939).

Less social than other small "peep" sandpipers in the genus *Calidris*, Baird's shuns mixed flocks and generally favors higher, drier inland habitats during migration, being fond of short-grass prairies, grazed land, wet fields, and lake margins. Feeding singly or in small to medium-large groups, the birds subsist almost exclusively on insects and spiders they acquire by pecking more than probing. Baird's typically forages on mud or very shallow water, shunning the deeper water habitats used by other shorebirds.

Feeding birds sometimes run and pause in the manner of plovers, and at other times move deliberately, especially when foraging in very shallow water. Loose flocks of 100 or more may be present at shallow-water habitats in some drier western United States during fall migration, where birds show territorial behavior toward their "preferred" feeding spots.

Wintering in western and southern South America from the Bolivian Andes to Tierra del Fuego, Baird's favors wide grassy plains and slopes and grazed shorelines of drying ponds. These habitats are also used in migration, as are alpine lakeshores and freshwater wetlands (most notably Cheyenne Bottoms and Quivera NWRs in Kansas).

Because of the remote breeding and wintering locations, population numbers and trends for this species are incompletely known. The latest generally accepted population numbers are 300,000 birds (Morrison et al. 2006). Since low numbers are often recorded during migration surveys, population trends are unknown.

▶ Least Sandpiper
Calidris minutilla

BIOMETRICS 4¾–5½ inches long; wingspan: 13¼–14 inches; weight: 15–28 grams (0.5–1 ounce)

STRUCTURE Chunky body shape that tapers in the rear; short-tailed and short-winged; smaller-headed than other "peeps" with finer-tipped and slightly drooping bill; typically shows a crouched, horizontal posture when foraging

STATUS Common and widespread; most common "peep" at interior sites, but outnumbered by Semipalmated or Western Sandpipers along coasts.

- -

This smallest of sandpipers behaves like a feathered mouse, creeping low to the ground with bent legs along the muddy margins of puddles and ponds, while its larger relatives (Western and Semipalmated Sandpiper) more typically enrich themselves in the benthic resources beneath shallow standing water.

▷ The planet's smallest shorebird is something of a feathered mouse. Always crouched and mostly brownish (with rust highlights like this breeding adult from Cordova, Alaska in May), Least Sandpiper has yellow legs and fine-tipped drooping bill that help distinguish it from other "peep" sandpipers.

◁ This mixed-shorebird flock from Florida in January finds the tiny, brownish-backed Least Sandpipers marginalized in the higher, drier, less food-rich foreground—which is probably where they would be anyway with no other birds in the deep water. Can you identify the other birds present? Other birds in this photo include the much larger White Ibis and Ring-billed Gulls, with other shorebirds being Lesser Yellowlegs, Sanderling, and Dunlin.

On open mudflats, where it may occasionally join other small sandpipers, Least is typically in the minority and easily overlooked because of its browner, mud-colored upperparts and tiny size. However, at some key stopover areas, particularly interior locations, and in freshwater locations adjacent to or near oceans or the Gulf of Mexico, Least Sandpiper may be the most numerous shorebird migrant, especially in early spring and midsummer.

In s. New Jersey in spring, when hundreds of thousands of migrating shorebirds gather to feast on the bounty of horseshoe crab eggs, Least Sandpipers are typically outcompeted on the egg-rich beaches and forced into the muddy Back Bay channels, where they forage on free-floating eggs ferried in by the tide, as well as small invertebrates.

The breeding plumage of this species differs greatly from its dull, grayish-brown nonbreeding condition. Showing a mix of strongly marked black and rust upperparts, a bold rust cheek and crown, and gleaming white underparts with a brown bib, a full breeding-plumaged Least Sandpiper is quite an eyeful.

Having the widest and southernmost breeding distribution of any Nearctic *Calidris*, Least Sandpipers breed primarily in subarctic tundra and boreal forests across the extreme northern regions of North America from Alaska to Newfoundland. They nest in coastal wetlands, bogs, sedge meadows, and tussock heaths, and in sand dunes at the southern portions of their breeding range in Nova Scotia and British Columbia.

Spring migration takes place from mid-March to early June, and northbound adults depart wintering grounds mostly in April. Arriving in small flocks on the breeding grounds from early May (south) to early June (north),

depending on latitude, males precede females, and they show a high degree of fidelity to previously successful sites. Favored habitats are treeless bogs with muddy edges near fresh and brackish ponds, muskeg and sedge meadows, and drier uplands near mudflats or freshwater pools.

Scrapes, excavated by the male, are situated in short marsh grass in moist areas. In wetter areas, nests are placed on tussocks or in tufts of sedge. Females improve the nest by adding vegetative lining and deposit 3–4 eggs, which are incubated by both sexes for 19–23 days, though males take a primary role. Precocial young leave the nest within one day of hatching and are initially tended by both adults, though females abandon young and males after about one week. Chicks are able to forage on small insects within hours of hatching, and they can fly after 15 days.

Southbound migration of adults from the breeding grounds occurs across a broad front from mid-June (failed breeders) to late June and July (females before males). Juveniles fledge from late June to early August and begin their southbound migration within a week or two. The first juveniles arrive across s. Canada and n. United States in mid-July, with peak passage in mid to late August. They first arrive along the Gulf Coast in early August, and in South America by mid-August. Eastern populations make transoceanic migrations from New England and the Maritime Provinces to wintering areas in South America and the Lesser Antilles.

Winter range includes both coastal and interior sites from Washington State and Virginia south to Chile and Brazil. In migration, birds utilize a wide variety of muddy areas, such as lakeshores, riverbanks, and rain pools, and in coastal areas, primarily sheltered tidal pools and margins of tidal flats. They are typically seen in small flocks of 10–20 birds, though larger groups may gather in prime habitat.

While many Least Sandpipers migrate through the interior, some make a transoceanic flight from the Bay of Fundy to northern South America, covering the 2,000 miles in approximately 70 hours. To accomplish this feat, the birds have to increase their weight by at least 30% in body fat (Cooper 1994).

Owing to its broad-front and protracted fall migration as well as a surreptitious nature, Least Sandpiper makes population estimates difficult. Estimates of 700,000 birds were reported in 2006 (Morrison et al. 2006), but winter counts in Suriname of 50,000–100,000 birds attest to a sizable population that might exceed Morrison's estimates. Least Sandpiper numbers appear healthy, with migration surveys indicating a stable population. Evidence suggests that this species is one of the most abundant shorebirds in the Nearctic region.

Nestled into its cozy nest on the tundra ecotone near Churchill, Manitoba in June, a Least Sandpiper is well camouflaged and virtually invisible to hunting eyes.

◤ While foraging, a juvenile Least Sandpiper avoids sinking into the shallow water covered by algae mats by flapping its wings at Jamaica Bay Wildlife Refuge, NYC in August. "If only I had webbed feet." Note the brownish trailing edge to the underwings, unique to Least among all "peep" sandpipers.

◣ Least Sandpiper has one of the most colorful, attractive juvenile plumages of all shorebirds, as this left bird exhibits in August in NYC. Note the grayish legs with yellowish tint on this very young bird due to a lack of bare parts pigment. A nonbreeding bird with frayed feather edges at right (probably a 1st-summer bird that molted early) feeds on a jetty in NJ in September.

◢ A group of nonbreeding Least Sandpipers waits out the high tide with a larger Dunlin on higher ground in Titusville, Florida in November. This tiniest shorebird shows very little variation in size and bill size/shape compared with other "peep" sandpipers.

△ A bright juvenile Least Sandpiper (NY, Aug) shows a compact flight profile with longish, broad-based, tapered wings. A worn bird in late July shows a mix of mostly breeding feathers on the upper back along with a few nonbreeding ones, and neatly arranged, recently replaced nonbreeding wing coverts.

▽ This well-fed, robust White-rumped Sandpiper is a high Arctic breeder and as such, is often still migrating north in early June, like this breeding adult in Cape May, NJ. Note the orange patch on the lower bill base, present on many White-rumpeds.

▶ White-rumped Sandpiper
Calidris fusicollis

BIOMETRICS   6–6¾ inches long; wingspan: 16–18 inches; weight: 25–51 grams (0.9–1.8 ounces)
STRUCTURE   "Athletic" look with bulky chest, sleek body, and long wings; primaries project well past tertials and either reach tail tip (males) or extend well past tail tip (females); short legs and medium-length, fine-tipped bill; horizontal stance
STATUS   Fairly common but easily overlooked in large shorebird flocks; most migrate through Great Plains in spring; Atlantic Coast in fall; population numbers appear stable

White-rumped Sandpiper is a graceful, long-winged shorebird that is slightly larger than the other "peep" sandpipers with which it forages. It has one of the longest migration routes of any North American bird, traveling in spring from s. South America to the high Canadian and Alaskan Arctic, a distance of 8,000+ miles. This journey is covered in several nonstop jumps that can last up to 60 hours and cover 2,500 miles and is made possible by their very long wings and extensive fat reserves acquired during important extended migratory stopovers.

Departure from wintering areas takes place as early as mid-February, but peak departure is from late March to late April. Some birds head to staging areas in n. South America, while others continue straight to the se. United States, where they add crucial fat reserves during extended stays. Most spring migrants pass through the interior of the United States east of the Rocky Mountains, with smaller numbers along the Atlantic Coast.

While most migrant Arctic breeding shorebirds leave the lower 48 states by mid to late May, White-rumped often remains a few weeks longer, until early June, as this high Arctic breeder has an exacting migratory timetable calibrated to meet the privations of their destination. In the high Arctic, snow and ice may still lie thick on the low marshy tundra favored by White-rumped, so they typically arrive on the tundra and start nest initiation in

Compared with Semipalmated Sandpiper in NJ in May (*right*), White-rumped Sandpiper is a chesty mesomorph among the "peep" sandpipers. Note White-rumped's streaked flanks with chevrons, wingtips that project beyond the tail, thicker neck, larger head, and noticeably larger size.

mid-June, later than most shorebirds other than some Pectoral Sandpipers.

Both sexes arrive simultaneously and unpaired, and males establish territories with energetic and vocal aerial displays that continue only through egg laying. Egg dates range from early June to early August, and mating is polygynous, with males often attempting to mate with as many females as they can attract to their territories with animated courtship displays. Once all the females have finished their egg laying in his territory, the male usually departs to stage with other males before migrating, leaving the females alone to incubate the eggs and raise the young. Peak hatching is late July.

Nests of leaves, moss, and lichen are constructed only by the female and are situated near lakes and streams in a well-concealed scrape, into which the female deposits 4 eggs that she alone incubates for 21–22 days. Females lead the precocial chicks to wetter, prime feeding areas as far away as a mile from the nest site, but some broods already close to prime feeding areas may remain in the natal area. Chicks can fly strongly in 16–17 days and are capable of great independence, giving females the latitude to join males at migratory staging areas in coastal areas.

Fall migration from the high Arctic to s. South America and the Falkland Islands takes place over a wide period from mid-July to December. Adult males depart in early to mid-August, followed by adult females in late August. Failed breeders may depart as early as mid-July. Most birds head to staging areas at James Bay and along the Atlantic Coast before heading to wintering grounds in

A breeding White-rumped Sandpiper from the high Arctic in Point Barrow, Alaska in June performs a single-wing raise to attract females to his chosen territory, or to warn other White-rumps to stay far from his domain. Even at the late date of June 17, there is still snow cover in Point Barrow, and the reason White-rumpeds migrate so late in spring.

South America on a long trans-Atlantic flight (O'Brien et al. 2006). From the n. South American staging areas, birds continue southeast along the coast until reaching the mouth of the Amazon River, where they turn inland and

△ A burly White-rumped Sandpiper molts into nonbreeding plumage in August in NYC. The short wings that extend only slightly past the tail and tertials suggest a male. A nonbreeding bird at right in October has mostly lead-gray plumage with gray streaks and chevrons on the underparts.

△ A colorful juvenile White-rumped Sandpiper shown with pale-breasted Western Sandpiper in late September in NJ is engaged in one of the most challenging migrations attempted by birds, one that carries them from the high Arctic of Canada and Alaska to s. South America, a distance of roughly 7,500 miles typically completed in up to 2,500-mile increments.

◁ White-rumped Sandpipers often submerge their heads to access food in deeper water substrates. When doing so, their tails point upward, as with this well-nourished, husky bird from NYC in August. Without sight, their prehensile bill tips feel and grip invertebrates in the mud below.

In characteristic foraging style, with head lowered and tail raised due in part to their long lower legs, this White-rumped Sandpiper has secured an insect larva morsel in a soft, muddy border at Jamaica Bay Wildlife Refuge in NYC in August. This profile with tail raised allows this species to be recognized in large flocks of foraging small sandpipers.

then south to wintering grounds from s. Brazil to Tierra del Fuego. This journey alone may take a month or more.

Juveniles depart breeding areas by about mid to late September and head primarily to staging areas at James Bay and along the n. Atlantic Coast before heading to the northern coast of South America by October/November. Arrival at their wintering areas in s. South America occurs a month or so later.

In migration and during winter, White-rumped prefers wet grassy areas, such as flooded rice fields, grassy margins of pools, inland lagoons, and grassy margins of mudflats. Occasionally they forage on open beaches and mudflats. They are sometimes seen in the company of Pectoral Sandpipers, and in migration, they often associate with Semipalmated Sandpiper (see page 191, upper photo). White-rumped often forages by submerging its head under the water, tipping its tail upward toward the sky. This profile is easy to see in flocks of Semipalmated Sandpipers, with the tail of the White-rumped rising up above the pack.

White-rumpeds feed mostly by probing in search of mollusks, aquatic worms, insect larvae, and plant material, often in fairly deep water, and birds may defend winter territories. The somewhat similar Baird's Sandpiper prefers drier habitat and picks more than it probes.

Population estimates for this species include a few segregated sites that contain anywhere from 250,000 to 500,000 birds, culminating in a world population number of about 1.7 million. Fall migration surveys indicate a significant population increase (3% per year) from 1974 to 2009, although trends in the last decade have been (non-significantly) negative. Overall, the population appears stable (Andres et al. 2012)

Buff-breasted Sandpiper
Calidris subruficollis

BIOMETRICS 7¼–8 inches long; wingspan: 17¼–18¾ inches; weight: 43–94 grams (1.5–3.3 ounces)
STRUCTURE Plump body with attenuated rear; small, squarish head, thin neck and short, straight bill with fine tip; upright stance
STATUS Uncommon migrant through Great Plains, though sometimes in large flocks; small numbers (usually juveniles) reach Atlantic and Northwest coasts in fall

This attractive sandpiper with warm tones, neat streaking, and wide-eyed appearance is one of the most delicately beautiful of all shorebirds. Its gentle nature away from the breeding grounds is soon put to rest in summer as males compete in a lek mating system that would elicit admiration from the best of carnival barkers. The elaborate single- and double-wing courtship displays are some of the most unusual in the avian world, and the male exhibits strong polygamous breeding behavior.

Besides being Kevin's favorite bird, long before he worked with them as a biologist in the Alaskan Arctic, Buff-breasted is one of the most sought after shorebirds by all birdwatchers. Adding to their mystique are their relatively low numbers due mainly to the indiscriminate slaughter of millions of Buff-breasts in the prairies of the United States by gunners, who in the 1800s shipped countless bodies of this bird in pickle barrels to the East Coast as a culinary delicacy.

Once thought to be a plover because of its superficial similarity to that family and its frequent association with American Golden-Plover at grassy feeding locations during

These winsome grasspipers were once slaughtered by the hundreds of thousands for food by market gunners during their spring migration through the prairies, a carnage from which they have not fully recovered. Some juveniles migrate along the Atlantic Seaboard in August and September, as this bird from Jones Beach, NY is doing. Buff-breast's high-stepping, herky-jerky gait is captured in this image.

migration, Buff-breasted Sandpiper is a medium-sized, dovelike shorebird that eschews open shallow-water locations used by most other shorebirds for its preferred feeding areas of short-grass habitats, including pastures, sod farms, golf courses, and meadows. It also favors wet rice fields, recently plowed or mowed agricultural areas, lakeshores, and occasionally sandy beaches, especially near sparse vegetation or wrack lines.

Buff-breasted forages with a peculiar pigeon-like head-bobbing motion and walks deliberately while picking food from the ground or foliage. Its diet is mostly terrestrial invertebrates, especially insects and spiders, and occasionally seeds. It is typically seen singly or in small groups but may gather in large flocks of hundreds or thousands at important spring staging areas.

A unique shot of a Buff-breasted's double-wing embrace display in Alaska warrants a second showing in our book. This posture is assumed after females are attracted to the single-wing display (see page 54). Multiple females often approach the male to vie for mating rights, after which he assumes no parental duties but continues to copulate with as many females as he can attract to his small mating post in June on the high Arctic tundra of Alaska and Canada. Kevin once saw 8 females pecking at the male's chest during a late snow melt year in 1993.

△ Juveniles are arguably more visually engaging than adults with their scalloped upperparts and gentle, doe-eyed expression. Julian Hough captured this interesting feeding angle on a beachfront in Connecticut on October 24, late for this migrant.

Once found in great numbers, and yet another shorebird whose numbers once defied quantification, its abundance is hinted at in accounts like: "it used to exist in the millions and was slaughtered in uncountable numbers" (Bent 1962). But like those other species whose numbers defied quantification, Buff-breasts were systematically annihilated principally along their spring migration route through the middle of the continent. The birds shared a witches' brew of unfortunate traits.

First, they flew and foraged in large, tight flocks that repeatedly returned to assess the damage done to fallen flock members. So confiding were the birds that gunners wading into their ranks reported that rather than flushing and leaving, birds would simply overfly the flock and land on the far side, or part and then coalesce behind the gunners.

The birds were also superbly fat (like the Eskimo Curlew); so much so that grease would ooze out of the shot-perforated bodies. Buff-breasts escaped the Curlew's fate of extinction by the narrowest of margins, with Edward Howe Forbush, a noted Massachusetts ornithologist, writing in 1912, "Apparently it is on the way to extinction." The conversion of native prairie to agricultural land also contributed to the bird's decline.

Curiously, despite its abundance, Alexander Wilson appeared unaware of the species, and Audubon admits to "never having met with it." However, Thomas Nuttall reported that the bird was "not uncommon" in the Boston market (Audubon 1827). The fact that Audubon once lived in New Orleans, and that Buff-breasted Sandpipers massed on the Texas coast in spring, attests to how narrow and proscribed the birds' spring migration route through North America was and still is.

It is a route that begins with northbound adults departing wintering grounds by early February and following the north-south mountain ranges in South America into April. Birds bypass both coasts and arrive in Texas and Louisiana from mid-April to early May after flying over the Gulf of Mexico. From there they follow a narrow corridor north through the central United States and Canada en route to their breeding grounds in the high Arctic of Alaska and central Canada.

As North America's only true lekking shorebird, 2–10 males attract females with elaborate and energetic wing-flashing displays, with each having adjacent lekking territoires of about 150 square feet. These leks are situated on a raised portion of tundra adjacent to an oxbow stream, where prospective females see the male's flashy underwings at great distances. Similar to aggressive carnival barkers, each male performs his flashing wing displays with vigor, especially when an interested female comes near.

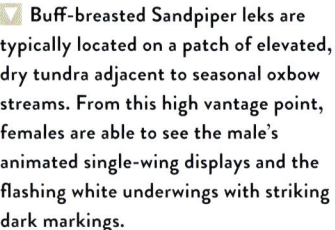

Even during migration, male Buff-breasted Sandpipers occasionally display to females, with their hormones raging even in early May. Scott Elowitz captured this scene in Nebraska, but copulation typically will not occur until high Arctic breeding grounds are reached in June.

Buff-breasted Sandpiper leks are typically located on a patch of elevated, dry tundra adjacent to seasonal oxbow streams. From this high vantage point, females are able to see the male's animated single-wing displays and the flashing white underwings with striking dark markings.

General behavior at the lek involves a great deal of standing around doing nothing, or manicuring the "mating post," a roughly 18-inch square piece of cropped tundra where the male will exclusively mate with a chosen female. Once a female approaches, the single-wing display switches to a cloaking double-wing display in which the male stands in his mating post and slowly rotates on his toes, all while making soft clucking sounds (see page 194, bottom photo). This display always reminded Kevin of the cloaking behavior performed by Dracula as he enveloped his hypnotized females.

Once a female, or multiple females, are attracted to this display (Kevin once saw eight females vying for the attention of a single male during a late snow melt summer season), the male will choose his mate and copulate with her after she lays prostrate on the tundra with her wings spread. After copulation, the female departs to lay her eggs in a nest that only she constructs, and the male starts displaying again to attract more females to his mating post.

An interesting sidebar to this whole process is that males from adjoining leks often pretend that they are interested females when they see true females approaching competing males, and they sneak in and break up the nuptial embrace by giving alarm calls. This results in both males aggressively flying high into the sky together with fluttering wings, and then eventually spiraling back to the ground and starting the process all over again.

It is a small wonder that successful copulation ever takes place, given the competitive behavior exhibited by competing males in the lek system. Despite the behavior of the males in choosing many mates while offering no support to the nesting process, females often even the

A hormonally charged **Buff-breasted Sandpiper** performs a double-wing embrace display at his mating post in the high Arctic in June near Kuparuk, Alaska (left). Rotating slowly with his chest inflated and making soft ticking sounds, this male is priming himself for an onslaught of females. A pseudo mating scene with all male birds (right) takes place near the end of the active mating season in late June when most females are sitting tight on their nests but male hormones are still elevated.

Competing male Buff-breasted Sandpipers often pose as interested females and break up ongoing pre-copulation displays at adjoining leks, and then explode into flight with alarm calls that suggest impending danger. When this happens, the affected male chases the intruder into high, stylized flight, after which both return to their leks and begin to display again. These displays have rarely been photographed, as Kevin worked with this species for four summers as a biologist on Alaska's North Slope and viewed this behavior only a few times. AK, June

score by copulating with multiple males to create their clutch of 3–4 eggs, thus strengthening the genetic gene pool (Richard Lanctot, pers. comm.).

Males have no role in the nesting process, and typically they leave the breeding grounds from late June to early July, when females are nesting. As prospective female numbers thin, some males may display to other males who are attracted to all the action and even attempt to mate with these nondisplaying males who did not set up leks of their own (see photo above).

Females do not build nests near the male's lekking territories, but they often situate their nests near several aggressive nest protectors, especially Black-bellied Plover and Ruddy Turnstone. Since the female has to leave her nest to feed, these other shorebirds help to deter predators that might threaten the nests in that area.

The 3–4 eggs hatch in 20–23 days, and chicks leave the nest in about 12 hours, after which they are fed by the female for several days. The female loosely attends the young chicks for up to 3 weeks, although unlike many

Once mated, often by several different males, female Buff-breasted Sandpipers retreat to a nest they built to lay and incubate the eggs and raise young alone. Tucked into tundra grasses, the feng shui of this nest is apparent, and the incubating female appears soft and serene. The eggs are subtly beautiful and the buckwheat leaves and Thamnolia lichen strands lining the nest are a reflection of her attention to detail. ALASKA, JUNE

An adult Buff-breasted Sandpiper walks on the tundra in Alaska in June and exudes the species' dovelike appearance. Two juveniles in flight at Jones Beach, NY in September show the characteristic scaly-backed look of juveniles. Long, tapered wings allow this species to migrate from high Arctic breeding grounds to s. South America, a distance of more than 8,000 miles.

sandpipers, Buff-breasted does not shift toward wetter environments, remaining true to the "upland species guild" shared with Golden-Plover and Baird's Sandpiper. Chicks feed mostly on insects and spiders they secure themselves, a dietary focus that carries over into migration and on winter territory in the Pampas grasslands of south-central South America.

Fall migration runs from mid-June to late October and is less constricted than in spring. Many birds still use a mid-continental route while others (mostly juveniles) migrate down the Atlantic Seaboard. Some birds apparently make a transoceanic jump from e. Canada directly to South America. On the Pampas of South America, Buff-breasted seeks out grazed grasslands, often in the company of American Golden-Plover and Upland Sandpiper.

While the global population declined from 1970 to the 2000s to about 30,000 birds (Lanctot et al. 2010 in Andres et al. 2012), recent recommendations from Lanctot suggest that a more accurate number is around 56,000. Some of this is based on migration counts from the Rainwater Basins in Nebraska (Jorgensen et al. 2008 in Andres et al. 2012). The higher value is closer to the numbers in spring counts on the Texas Gulf Coast (Morrison et al. 2006). Few long-term datasets exist to assess population trends.

▶ Pectoral Sandpiper
Calidris melanotos

BIOMETRICS 7¾–9¼ inches long; wingspan: 16¾–19½ inches; weight: 50–117 grams (1.8–4.2 ounces)

STRUCTURE Bulky and broad-chested, but attenuated body shape, especially in rear; long, broad wings with tapered tips; short to medium length bill that droops near tip; horizontal feeding posture; primaries equal to tail tip

STATUS Fairly common migrant through Great Plains in spring and fall; fairly common along Atlantic Coast in fall; less common elsewhere; population trends indicate higher numbers in the 1980s, thus probably declining slightly

------

This chunky, chesty, corduroy-bibbed *Calidris* sandpiper is almost as large as Red Knot, and almost three times the weight (males) of Least Sandpiper. But superior size is just the beginning of a series of peculiarities that sets Pectoral Sandpiper apart from its congeners, beginning with socialization. Rarely does Pectoral Sandpiper join mixed-species flocks (but it will if freshwater habitats are scarce due to drought), nor are you likely to find Pectoral Sandpiper on sandy beaches or open mudflats. This is a bird of moist tundra, flooded fields, wet grassy meadows, lake edges, and the lushly vegetated edge of impoundments.

While the flight calls of other sandpipers are a high-pitched chip, trill, or squeak, Pectoral "burps." Its repetitive territorial display call given in flight is a deep, low hooting reminiscent of air blown over a bottle top and is an important component of courtship and territorial displays that are histrionic bordering on the buffoonish. Indeed, the male Pectoral Sandpiper puts a good deal of its procreative

▶ Named for the dark, streaked pectoral bib that males inflate with internal air sacs during courtship, these buffoonish birds breed from e. Siberia and Alaska across much of Canada in high Arctic habitats. A promiscuous species, males attract harems of females to their territories and mate with all of them but assume no parental duties. ALASKA'S NORTH SLOPE, JUNE

◀ Larger than "peep" sandpipers, Pectoral Sandpipers rarely forage on open mudflats, preferring instead to feed in wet, well-vegetated freshwater marshes, meadows, and lake/pond edges. This male in breeding plumage in early April at High Island, Texas appears in typical shallow, freshwater habitat.

▲ Male Pectoral Sandpipers conduct low, aerial flight sorties to define and reinforce their fairly large harem territories. Fatty air sacs in his chest are inflated during these flights while he makes a series of deep whooping sounds that increase in volume near the end. He also struts around females while bowing and making these same sounds. Could any prospective female doubt that this is the biggest, baddest, most virile Pectoral on the tundra? ALASKA, JUNE

energy into defending its large territory and wooing multiple mates to his harem. Unlike other members of the genus where males play an important, if not dominant role in incubation and brood rearing, male Pectorals begin and end their parenting role with copulation. Next, please!

Males set up large territories amid huge expanses of similar moist tundra habitats, and each male attempts to attract as many females as possible to join his harem. He is a promiscuous shorebird in this respect. After copulating, each female lays her eggs in a nest that she builds within his territory, and she performs all the nesting duties without his help. The male, however, defends the entire territory vigorously from other male Pectorals and predators, and he may watch over the territory during the incubation stage as well.

Females may place their eggs outside the male's territory and consort with other males, especially if the initial male has abandoned the territory. By the time eggs hatch, most males are migrating south, and these movements are usually through the midcontinent and typically occur in July. The difference in size between larger males and females is pronounced and consistent with a promiscuous mating system. One of Kevin's study plots in the Alaskan Arctic in 1992 included a Pectoral's territory, and the male weighed 117 grams, while a nearby female weighed only 57 grams, less than half his weight. Much of this extra weight is concentrated around the inflatable fatty chest sac of the male.

Breeding mostly in moist graminoid (grasslike) tundra in both Russian and North American Arctic locations, Pectorals occupy similar grassy habitats in South America in winter after making an 8,600-mile flight from North American breeding areas. This journey is even farther if you herald from Russia and travel to Alaska before turning south. Some individuals also winter in the South Pacific and Australia.

▼ A male Pectoral Sandpiper with pectoral sacs inflated surveys his harem from an elevated ridge on the tundra in Prudhoe Bay, Alaska in June. Even after copulation with all their females is over, males often remain in their territories and offer protection by alerting incubating females to potential danger, and by harassing aerial predators.

Pectoral Sandpiper has the distinction of being one of the longest-distance shorebird migrants in North America.

Spring migration takes place between late February and late June, with most birds departing wintering grounds in March and arriving along the Gulf Coast in early to mid-March, with peak passage there between late March and late April. Most Pectorals migrate north through the Great Plains from mid-April to late May, with a peak in mid-May. Cheyenne Bottoms and Quivera NWRs might host half the world's population during this time. Arrival on the breeding grounds takes place from late May to mid-June, with a later arrival for some birds in late June to early July (probably 1st-year birds).

Nest initiation is primarily early to late June, and incubation is 21–23 days. All the eggs hatch within a few hours of each other, and the young are only helpless as they stretch their wings and legs and allow their blood to circulate through their tiny bodies. Young can fly after about 23–25 days. Males depart breeding areas from early to mid-July; females from mid-July to mid-August; and juveniles in mid-August (earlier from some high-latitude locations, especially Siberia and Alaska's North Slope). Some of these juveniles disperse to the East and West coasts.

Fall migration is a drawn-out affair, lasting from late June for failed breeders and from early July (males) to early December (juveniles). Peak numbers pass through the n. Pacific Coast and n. Plains from mid-July to mid-August, and from the s. Pacific, Gulf, and mid-Atlantic coasts in August. Adults are rare in North America after September. Arrival at wintering sites in South America takes place mostly from mid-September (adults) to late October (juveniles).

Previous estimates of Pectoral's North American population were listed at 500,000 birds (Morrison et al. 2006), but more recent Arctic PRISM surveys (Bart and Smith 2012 in Andres et al. 2012) revised this number upward to 1.6 million. This dramatic increase was based on several factors, including a US Prairie Pothole survey in 2008

A small Pectoral Sandpiper, probably a female, shares a muddy freshwater pool with Greater Yellowlegs and a Baird's Sandpiper in retarded plumage in early April in Galveston, Texas.

A Pectoral Sandpiper nest with two hatched chicks and another egg pipped is shown from Kuparuk, Alaska in early July. Chicks are helpless for a few hours after hatching while their blood pumps to awaken muscles, wings, and body. A few-days-old chick at right sports attractive natal plumage, with CUTE in capital letters. ALASKA, EARLY JULY

Juvenile Pectoral Sandpipers in flight in September show the very long, tapered wings that enable them to complete one of the longest migrations of all shorebirds from the high Arctic to s. South America. An adult at right is beginning its molt to nonbreeding plumage in late August in NJ, and its prehensile bill with flexible tips is shown.

that estimated a southward migration population of 713,424 birds (Skagen et al. 2008 in Andres et al. 2012). Given the wide range of migration of this species, this number was a starting point for higher population estimates. While this number appears high, Edward Howe Forbush (1912) accounted the "Grass-bird," as it was known, "one of the most numerous species … and it was much sought for the market."

Sharp, natty juvenile plumage is shown on this migrant bird picking insects off low vegetation at Jamaica Bay Wildlife Refuge in NYC in mid-September. This plumage is a mix of attractive earth tones with black, white, and brown shading highlighted by a buff background.

Semipalmated Sandpiper
Calidris pusilla

BIOMETRICS 5½–6 inches long; wingspan: 13–14 inches; weight: 14–41 grams (0.5–1.4 ounces)

STRUCTURE Plump body with even weight distribution; straight, blunt-tipped to drooping fine-tipped bill; short wings that don't extend past tail at rest; females average longer-winged and longer-bodied than males

STATUS Locally abundant; most abundant "peep" at Atlantic coastal sites; uncommon inland; somewhat rare in West; a few winter in s. Florida. Population numbers have dropped steadily over the last several decades from about 4 million birds to the present estimate of 2.26 million, with much of this decline attributed to the eastern Canadian breeding population.

What this small sandpiper lacks in size, it gains in stature with its extremely feisty territorial behavior both on and away from the breeding grounds. It is not unusual to see Semipalmateds attacking nearby Semis and other small sandpipers that make the mistake of straying into its "chosen feeding territory." It might amount to only a small 3 square feet of mudflat, but it might as well be the last stand of the Alamo given the ferocity of the attack.

Semis threaten other small sandpipers with a theatrical display of fluffed-up neck feathers, raised tails, drooped wings, and tilted bodies, which is often enough to drive them away. Lacking arms to fight with, sandpipers try to land on the back of the intruder, where they can inflict damage to the neck and head. With a Latin species name *pusilla*, which means "miniature" or "very small," this feisty sandpiper certainly does not adhere to its label but instead acts with a big spirit and courageous heart.

Any contemplation of this small sandpiper invariably invokes numbers, flocks at the very least, but more typically massed concentrations of migrating birds in excess of 100,000. Indeed, vast numbers are a Semipalmated Sandpiper hallmark, and with this species, we gain a glimpse into the pre-market gunning heyday of other shorebird populations.

At key migratory staging areas and on the tidal flats of n. South America, where Semis winter, flocks in excess of 300,000 birds were fairly common as recently as the 1990s, with today's population numbers almost halved since then, from about 4 million birds worldwide in 1990 to about 2.4 million today. Outside the Arctic breeding season (June–August) when monogamous pairs maintain territories, it is rare to encounter a solitary Semipalmated Sandpiper. In the absence of other Semipalmated Sandpipers, this obligate flocking species may join Sanderling, Dunlin, and Western Sandpipers.

Pete recalls only a single occasion when he saw a lone Semipalmated Sandpiper. It was at the end of a landing strip serving an abandoned DEW Line site appropriately named "Lonely." (In the high Arctic, DEW [Defense Early Warning] sites were US Air Force facilities used to monitor any bomber activity from Russia after WWII.) The bird was incubating eggs on a pile of sand, 40 feet from the Arctic Sea and surrounded by tussock tundra. Both sexes share in incubation duties, with each parent taking 3–5-hour shifts, and Pete's solo Semipalmated surely had a mate somewhere nearby.

With touches of rust and ginger enhancing their breeding plumage, Semipalmated Sandpipers fall short of flamboyant, but this endearing "peep" does not lack for charm. Kevin considers this species the most variable of all shorebirds with respect to size, bill size and shape, and plumage, as these two breeding birds can attest in late May in NJ.

Sandpipers and Allies (Family Scolopacidae)

▶ Semipalmated Sandpipers are some of the most aggressive shorebirds of all when it comes to defending feeding territories against other small sandpipers, even during migration. These two adult birds are squaring off over a small feeding patch of pond border, and the one in the air is trying to land on the back of the other to get the physical advantage in this squabble. NYC, LATE AUGUST

▼ At key staging areas during migration, concentrations of Semipalmated Sandpipers may number in the hundreds of thousands. This group of 60,000 birds, with a few Dunlin mixed in, staged in Heislerville, NJ in May, where both tidal flats and impoundment pools provide ample food for fattening up before their final push to Arctic breeding grounds.

▼ A breeding Semipalmated Sandpiper in flight in May shows a compact body shape and moderately long wings. A lone bird molting into breeding plumage in April in Texas forages with three Western Sandpipers at right. Besides plumage differences, Semipalmated has a more slender, more evenly balanced body shape and a smaller, rounder head compared with Western's bulky, front-heavy body structure and larger, deeper head that squares off in the rear.

In spring, northbound migrants depart wintering grounds mostly in April, with peak numbers arriving along the Gulf Coast in late April and early May, and in mid to late May into early June in the Northeast. Most breeding areas are reached by late May and early June, with males preceding females by several days.

After the birds arrive in Arctic breeding locations, copulation takes place after males establish territories, and females lay 3–4 eggs in 3–4 days, a taxing and extraordinary physical feat. Females have to feed continuously during this time to accumulate the necessary fat and body weight to achieve this task. Semipalmateds build their tundra nests in moist sedge habitats, and the grasses grow up and around the incubating birds, making them difficult to find. Both adults incubate the eggs, which take 20–22 days to hatch.

Like most other sandpipers, young Semipalmateds are strongly precocial, able to feed themselves soon after hatching. Both adults attend and brood young after hatching for up to 2 weeks, though females typically abandon young and male within 11 days. Brooding of the downy chicks is necessary both to protect them from an array of predators and to keep chicks warm when ice fog drifts in off the ice cap, dropping temperatures below freezing.

Young are able to fly after 16–19 days if they avoid marauding jaegers, Arctic and Red foxes, and other aerial predators. After fledging, they follow adults southward, joining the ranks of birds using routes prescribed by their breeding location. Southbound adults begin to depart breeding areas by mid-July (females before males), with failed breeders heading south by late June. Arrivals begin in South America by mid-July, and most are gone from North America by early October. Juveniles fledge by mid to late July and begin their southbound migration within a week or two. Most are gone from North America by mid-November.

But since juveniles migrate several weeks after adults depart, they have to chart this several-thousand-mile journey without adult guidance or previous experience. This information is programmed into their brains that are the size of a pea, and they not only migrate along the same routes as adults but also arrive at the same migratory locations and wintering grounds more than 2,000 miles away. This imprinted migratory guidance is similar to our modern-day GPS technology in both accuracy and accessibility and one of the true mysteries of shorebird migration.

The breeding range of Semipalmated Sandpiper is wide ranging but relatively narrow latitudinally, stretching from Alaska to the Atlantic Maritimes, where they nest in subarctic to Arctic habitats. Birds from eastern and central Canada migrate down the Atlantic Coast in late summer and fall, with many birds making the long transoceanic migration from New England and the Maritime Provinces to wintering areas in eastern portions of South America, a flight of up to 2,000 miles. Western Arctic breeders migrate primarily through the prairies to wintering areas in South America.

The high Arctic breeding season is short, so birds need to work fast to successfully nest and raise young. The mating Semipalmated Sandpipers at left show a huge female that has put on large amounts of body fat to allow her to lay 4 eggs in 4 days, an amazing feat in any playbook! A Semi on its tundra nest is mostly hidden by grass grown up around the nest, making the bird very difficult to see by predators and biologists. ALASKA, JUNE

△ Given the perils of the high Arctic, these three recently hatched Semipalmated Sandpiper chicks in June have a long path to fledging. One brood per season is typical, except for re-nests after failure, and both adults attend young that leave the nest soon after hatching. A crisp, colorful juvenile Semipalmated migrant is shown from Cape May, NJ in mid-August, less than two months after hatching and well over a thousand miles from breeding grounds in Arctic Canada.

By whatever route, their destination is coastal South America, where birds mass on intertidal flats and fan out like a hungry wave, pecking and probing rapidly as they move. Always on the move and surging forward in response to the falling tides, feeding birds make mudflats vibrate like Brownian motion viewed through an electron microscope.

Watching the hungry wave of birds pass, it seems a marvel that any small marine invertebrates can survive the onslaught, but while these sandpipers are opportunistic, they can also be surprisingly choosy about prey. In Suriname, one of their main wintering areas, they feed mostly on small crustaceans. In Delaware Bay in May, northbound birds feed almost exclusively on the lipid-rich eggs of horseshoe crabs. Semipalmated Sandpipers around Delaware Bay in spring number up to 100,000 or more birds, and Kevin once counted a single flock of 60,000 Semipalmateds in May at Heislerville, NJ. During the breeding season, assorted insect larvae and spiders constitute important prey items.

◁ As aggressive defenders of feeding territories, in this case a small patch of mud, Semipalmated Sandpipers warn other small sandpipers not to get too close by raising their rear bodies and lowering their heads in a menacing fashion while feeding, as this adult is doing in August in NYC.

Semipalmated is one of the most variable shorebirds in the world with respect to sexual dimorphism (differences between males and females). Bill size ranges 15–25 mm, with males having shorter, straighter, stubbier bills from 15 to 20 mm and females showing longer, finer-tipped and often drooping bills from 20 to 25 mm. Bill size and shape is clinal from west to east, with some Eastern breeding Semis from the Ungava region of Quebec Province having long, fine-tipped, drooping bills up to 26 mm in length, which is longer, more fine-tipped and drooping than some male Western Sandpipers (KK, pers. comm.). The weight of Semipalmated varies greatly from 14 to 41 grams, with the larger measurement reflecting very large, well-nourished females and the lower number, undernourished males.

There are three recognized breeding populations of Semipalmated Sandpiper, grouped in part by noted differences in average bill lengths: western (Alaska); central (w. Canadian Arctic); and eastern (e. Canadian Arctic). While the Alaskan and central populations seem relatively stable since the 1970s, the eastern population has declined greatly since this time but may have reversed this decline in recent years (Andres et al. 2012). One of the reasons for the decline of eastern breeding Semis is the indiscriminate netting and hunting of these small birds for food in South America by humans.

The total population estimate is around 2.26 million birds (Andres et al. 2012), making Semipalmated Sandpiper one of our most abundant shorebirds. But as the loss of the Passenger Pigeon demonstrates, large numbers are no guarantee against extinction. Eastern breeding birds that stage in Delaware Bay are particularly vulnerable to oil spills and to reductions in horseshoe crab egg numbers.

On the Delaware Bayshore in NJ, mostly in the last half of May, tens of thousands of Semipalmated Sandpipers outnumber Red Knots and Ruddy Turnstones while feasting on horseshoe crab eggs.

A juvenile Semipalmated Sandpiper with a short, stubby bill strikes an angelic pose as he balances himself by flapping his wings while feeding on floating mats of algae. This eye-level image in August, NYC shows the long tapered wings that allow this species to migrate long distances, with the reflection adding to the artistic value of the photo.

▶ Western Sandpiper
Calidris mauri

BIOMETRICS 5½–6¾ inches long; wingspan: 14–15 inches; weight: 22–35 grams (0.8–1.2 ounces)
STRUCTURE Front-heavy weight distribution with thick neck and larger, deeper head than Semipalmated that squares off in the rear; most have longer, often more fine-tipped bill than Semipalmated; typically stands more upright than Semi at rest due to front-heavy weight distribution.
STATUS Abundant, especially at many Pacific Coast locations; less common in Northeast; uncommon in continental interior except during migration.

Decked out in its bold and flashy breeding plumage, this large "peep" is a stark contrast in winter with its dull gray and white nonbreeding garb. With rufous and gold markings on the head and upperparts, breeding adult Western Sandpipers are the most colorful of the tiny North American sandpipers known as "peep." Westerns show a high degree of sexual dimorphism with respect to size and structural features, with females slightly to much larger than males and having considerably longer legs and bills (up to 30%).

As possibly the most abundant shorebird in North America with an estimate of 3.5 million birds including Alaska and extreme e. Siberia (Andres et al. 2012), this species is one of the more common shorebirds in winter in North America, especially at western coastal locations. Flocks may number in the hundreds of thousands in California and Alaska in spring, representing one of North America's greatest wildlife spectacles as the birds lift up and wheel about in amazing synchrony.

In 1973, millions of Western Sandpipers were estimated at the Copper River Delta near Cordova, Alaska, where a majority of this species stage in early to mid-May. The spectacle here is truly amazing, with numerous flocks of Westerns spread out over the tidal mudflats. Here they fatten up on tiny macoma clams and mussel spat, a thin layer of mussel larvae biofilm that covers the surface of tidal mudflats. Since they pick and probe so quickly, only slow-motion film in the 1990s documented them feeding on this spat, and so did Kevin with his still camera (see page 21, upper photo).

In spring, northbound migrants depart wintering grounds in late March and April, with peak numbers along the Gulf Coast in early April, and through the central and n. Pacific Coast in late April. Western Sandpiper's diet is diverse and reflects the seasonal food abundance of appropriately sized prey at stopover locations. Most breeding areas are reached by mid to late May, with males arriving first. They establish territories on low Arctic

◁ One of the most strikingly patterned "peep," Western Sandpiper dons touches of rufous color and arrowhead markings on the underparts in breeding plumage. This male with a short, thick bill from Cordova, Alaska in May has gained an overabundance of body weight and fat to carry him to Arctic breeding grounds.

▶ Up to 4 million Western Sandpipers stage in Cordova, Alaska in early May prior to making a last jump to somewhat restricted Arctic breeding grounds in nw. coastal Alaska. This flock of several thousand birds, along with numbers of Dunlin, are flying in a tight murmuration to escape a Merlin that appeared on the Copper River Delta mudflats.

◀ While catching a few winks during a high tide respite, the appearance of a Merlin caused these roosting Western Sandpipers to come to high alert in Cordova, Alaska in May. You can see concern in their eyes with their crown feathers raised as they prepare for their exit flight

tundra near water and often construct their nests in a small patch of dry heath tundra among large expanses of wet grass-sedge tundra. Both males and females often return to the territories used the previous year.

Perhaps owing to the high density of breeding birds (much of the world population of Western Sandpipers breeds in a relatively small area in coastal northern and western Alaska), males are dogged defenders of their territory. Tail chases between rival males are commonplace until young hatch, and females appear to select in favor of males that mount a vigorous territorial defense. Egg laying usually begins in late May/early June, soon after snow melts. Incubation of eggs is 21 days, with both adults participating. The male, however, does most or all chick rearing. Young can fly after 17–18 days.

Southbound adults begin to depart breeding grounds by late June (females before males). Failed breeders may begin to head south slightly earlier. While the first fall migrants begin to appear in many areas across the United States in early July, peak numbers pass through in mid-August, gathering in flocks that may number in the multiple thousands. Important "fall" staging areas include the Yukon-Kuskokwim Delta in Alaska, where totals of 700,000 birds have been noted in this pre-migratory feeding area.

A female Western Sandpiper incubates her eggs in a cozy nest on the Y-K Delta in w. Alaska in late May. Nests are shallow depressions often placed on an elevated tussock beneath a stunted willow or birch. Both parents incubate the 3–4 eggs, but the male does most of the brood rearing. This photo would make a great jigsaw puzzle, with the white reindeer lichen enhancing the mood! This image by Ted Swem is one of the authors' favorite nest shots.

Western Sandpiper is more common along the Atlantic Coast in fall, with this species a relatively rare migrant in spring. First juveniles arrive in Pacific Northwest areas in late July and along the Atlantic Coast a few weeks later, with peak passage at both locations between late August and mid-September.

Restricted breeding range notwithstanding, this abundant shorebird floods much of the continent every summer as southbound migrants. At times and in places, this small peep is the most abundant migrating shorebird species, especially in Western locations. While many winter along the Pacific Coast from Washington State to Peru, a

You can almost hear the air sizzle with the wings of these fast-flying Western Sandpipers. Starting in early July and peaking in mid-August, hundreds of thousands of Western Sandpipers stage on the Yukon-Kuskokwim Delta in w. Alaska prior to migrating. ALASKA, MAY

Two female Western Sandpipers with long, drooping bills are shown here. A fresh juvenile (*left*) and nonbreeding adult (*right*) were both photographed at Stone Harbor, NJ in mid-September. Note the nonfunctional rear toes on both birds, which were probably used for perching in trees many years ago.

goodly number also winter coastally and inland from n. California east to Texas and Florida, and along the Atlantic Coast from New Jersey to Florida and south to n. South America and the Caribbean.

Unlike many long-distance migrants that make transoceanic crossings, Western Sandpiper appears not to lay down lavish fat reserves prior to migration. Their migratory strategy is to use shorter leaps and stopovers that are more frequent. In winter and during migration, Westerns favor sandy beaches and tidal mudflats, where they seek an abundance of insect larvae, crustaceans, and marine worms. During migration, which is primarily along the Pacific Coast with smaller numbers moving through the interior, this species mixes freely with the similar Semi-palmated Sandpiper and larger Dunlin.

Population estimates for this species remain at about 3.5 million (Morrison et al. 2006), with some migratory counts and Christmas Bird Counts suggesting a continued decline in numbers, although the reliability of these counts is low because of incomplete coverage (Andres et al. 2012). The oldest recorded Western Sandpiper was 9 years, 2 months when recaptured during banding operations in Kansas.

Western Sandpipers molting to breeding plumage are shown feeding in a tight flock at Bolivar Flats, Texas in early April. While not as aggressive as Semipalmated Sandpiper during migration, birds with tails up and wings out are warning their neighbors not to get too close to "their" feeding areas, lest trouble will ensue.

▶ Short-billed Dowitcher
Limnodromus griseus

BIOMETRICS 9¼–10 inches long; wingspan: 18–20½ inches; weight: 65–154 grams (2.3–5.5 ounces)

STRUCTURE Chunky but attenuated with even weight distribution in front of and behind legs; long bill with thick base and blunt to thin tip; often shows a kink near the bill tip; slimmer and flatter-backed on average compared with Long-billed, with a straighter undercarriage

STATUS Common in coastal areas; generally much scarcer inland than Long-billed, although midcontinental subspecies *L. g. hendersoni* occurs inland during migration. Population trends are considered stable for *L. g. hendersoni/griseus* and unknown for *L. g. caurinus* (Andres et al. 2012).

This evenly balanced shorebird with the Popsicle-stick bill is a common migrant typically seen foraging in tight-packed flocks, all the while probing for food in shallow water with metronome-like regularity. While other probing shorebirds range about while feeding, it is not uncommon to see a group of dowitchers feeding in the same location several hours later, unless displaced by a hunting raptor. They show little to no aggression to other dowitchers or other encroaching shorebirds while feeding, unlike other shorebirds that show aggressive behavior toward intruders.

The name "Short-billed" doesn't seem appropriate for this species, as the bill is quite long, even compared with Long-billed Dowitcher. Some female Short-billeds can have long bills about the same length as some male Long-billed Dowitchers (female sandpiper bills average from slightly to much longer than males). However, male Short-billed and female Long-billed dowitchers are more easily told apart by their bills, with no overlap in the bill length of these birds.

Hunted nearly to extinction in the 1800s, these once abundant, highly social, unwary birds were easily decoyed and would return to fly over wounded flock members until all were killed, or until market gunners ran out of shot and powder. Edward Howe Forbush, a prominent ornithologist, wrote in 1912 that he believed that without an end to spring gunning and "overshooting," the species "will join the Dodo and Great Auk" to extinction. Happily, protection was afforded in time, and today Short-billed Dowitcher is much recovered. It is not uncommon to see hundreds of these birds feeding shoulder to shoulder at key migratory locations.

Despite its abundance, this species was dogged for many years by confusion linked mostly with uncertainty relating to its relationship to Long-billed Dowitcher. This confusion was not rectified until 1950, when the determination was made that Short-billed and Long-billed Dowitcher were indeed separate species, as first proposed by scientists in 1932. The reason for this confusion is that both species share many overlapping traits, but recent information in *The Shorebird Guide* (O'Brien et al. 2006) shares a number of field characters, including physical

◀ While few would call the bill of Short-billed Dowitcher "short," the Popsicle-stick bill of these pear-shaped shorebirds averages slightly to much shorter than the closely related Long-billed Dowitcher. Breeding plumage of this *L. g. hendersoni* is the brightest of all three Short-billed subspecies. CHURCHILL, MANITOBA, JUNE

◀ Direct comparison of a male *L. g. griseus* Short-billed Dowitcher (right) and a female Long-billed Dowitcher (left) shows the larger size, longer bill and legs, and more front-heavy weight distribution of Long-billed in this photo from NYC in early August. Note the mostly horizontal backs on all the nearby Short-billeds compared with the steep back angle on the Long-billed as she balances her front-heavy weight distribution.

△ Short-billed Dowitchers form small to large flocks during migration, and this silhouette shot of a feeding group in NJ in May shows the loosely spaced dynamics. Can you identify a few other shorebirds in this photo? Other shorebirds include 2 Black-bellied Plovers, 9 Semipalmated Plovers, and 1 sneaky, smaller Dunlin facing away right of center.

structure and bill shape, that can be used to separate them, even in nonbreeding and juvenile plumages.

Three subspecies of Short-billed Dowitcher are recognized today, all heralding from different taiga and subarctic breeding locations.

■ *L. g. caurinus*: breeds along the south coast of Alaska and Yukon Territory and winters along the Pacific Coast from Washington to Peru.
■ *L. g. hendersoni*: breeds in central Canada from e. British Columbia to Manitoba and probably winters along the s. Atlantic and Gulf coasts and along both coasts of Central America.

■ *L. g. griseus*: breeds in Quebec and Newfoundland and probably winters mostly in the West Indies and coastal Venezuela and Brazil, although some may winter in the se. United States.

In summer, many Short-billed Dowitchers breed in the taiga shield ecotone (*L. g. hendersoni* and *L. g. griseus*), where northern trees such as spruce, tamarack, and birch become stunted or scant as the boreal forest gives way to Arctic tundra. They nest in wetlands, often near the edges of bogs (muskegs), small lakes, or wet meadows with willows and alders. Some also nest in river floodplains. In this taiga habitat, birds often sit high up in dead

spruce or tamarack trees to survey their breeding territory. The western subspecies *L. g. caurinus* breeds in subarctic habitats along the Alaska coast, often in bogs or wet meadows near open boreal forest habitat.

Spring migration takes place between March and early June, with different timing for the three subspecies. Many one-year-old birds oversummer on or near the wintering grounds, while others return partway or all the way north to the breeding grounds. Birds migrate during both day and night under a variety of weather conditions.

Birds arrive on territory unpaired in late May or early June, with males arriving first. Monogamous pairs situate their grass-lined nest in thick wet vegetation beneath concealing branches. Females often depart just before the eggs hatch, leaving precocial young in care of males for about 12–14 days. Male duties involve brooding, protection from predators, and leading young to food, which they acquire first by pecking. By the time young fledge in 16–17 days, they are probing like adults.

Fall migration takes place between late June and early November, with most adult females, failed breeders, and nonbreeders departing the breeding grounds in late June or early July, soon after egg hatching. These birds typically fly directly to coastal areas. Breeding males depart

A breeding male *L. g. hendersoni* Short-billed Dowitcher (very short bill) is surveying his breeding territory from the top of a stunted spruce in taiga habitat near Churchill, Manitoba in June. From this elevated perch, birds can spot any intruders or predators that enter their nest area.

In May, these Pacific race Short-billed Dowitchers (*L. g. caurinus*) are at the peak of breeding plumage as they feed near Homer, Alaska. This subspecies is very similar to Long-billed Dowitcher in plumage, and the most difficult to separate from Long-billed as well, especially in winter.

A very bright juvenile Short-billed Dowitcher, probably the *L. g. hendersoni* subspecies, probes the shallow water substrate with its prehensile bill tip at the Jamaica Bay Wildlife Refuge, NYC, in early September. These birds forage with a sewing-machine motion and don't wander much when feeding, unlike other species of probing sandpipers.

All dowitchers have a white slash up the back, as shown in this nonbreeding Short-billed Dowitcher in flight. A nonbreeding bird at right is taking a running start to lift into flight while showing the complex dark markings on its underwing. Short-billed Dowitchers favor brackish or saltwater tidal flats in winter, as opposed to Long-billed, which prefers freshwater habitats.

about 2 weeks later, after young have fledged. Juveniles depart the breeding grounds in early to mid-August and head primarily to coastal areas. Arrival at wintering sites continues through late October/early November.

In winter, the birds prefer tidal habitats but also inhabit mangroves and man-made ponds in the tropics. At inland locations during migration, freshwater wetlands, flooded fields, sod farms, and sewage ponds are used. Food sources include buried invertebrates, especially marine worms, mollusks (small clams), crustaceans (fiddler crabs, shrimp), isopods, and amphipods. Most food is captured and swallowed within the substrate, but marine worms are pulled to the surface. Breeding dowitchers feed on a great variety of insects as well as spiders and snails. Some of these are picked from the surface and plant leaves, contrary to their subsurface feeding style.

Population numbers are estimated at 75,000 for *L. g. caurinus* and a combined 78,000 for *L. g. hendersoni/griseus*, for a total of 158,000 Short-billed Dowitchers (Morrison et al. 2006). While migration counts from the 1990s to early 2000s showed declines, especially in Ontario, recent counts are similar to the early 1980s (Andres et al. 2012). Delaware Bay counts are migration counts, which can be unreliable for many reasons. The oldest Short-billed Dowitcher was at least 13 years, 11 months old when it was recaptured and rereleased in Delaware during banding operations.

Short-billed Dowitcher males and females of the subspecies *L. g. griseus* (males are the smaller ones) take a cat nap at high tide in early August in NYC. A juvenile Short-billed is at right. Molting Black-bellied Plover and Semipalmated Sandpiper (front right) share space with these dowitchers.

Long-billed Dowitcher
Limodromus scolopaceus

BIOMETRICS 9½–10½ inches long; wingspan: 18–20½ inches; weight: 90–114 grams (3.2–4 ounces)
STRUCTURE Chunky and with rounded back and belly; chesty, front-heavy weight distribution, unlike the evenly balanced Short-billed; long bill averages longer than Short-billed and has a thinner base with fairly even width and finer tip; females are especially rounded with exceptionally long bills that don't come close to overlapping with Short-billed.
STATUS Common in the South and West, less common in the Northeast (rare in spring); population trends are unknown due to variability in migration counts, with more Arctic PRISM counts necessary over the long term.

With a chesty, front-heavy weight distribution and rounder undercarriage than Short-billed Dowitcher, this longer-legged, longer-billed, more rotund dowitcher was first described by Thomas Say in 1823 during an expedition to the Rocky Mountains. The bird was collected near Council Bluffs, Iowa, and presented by Say as a new species, distinct from Short-billed Dowitcher, a differentiation actually made 4 million years earlier when the species diverged from a common ancestor. However, it took more than a century before Say's determination was accepted, and Long-billed Dowitcher was finally accorded full species rank by the American Ornithologists' Union in the 1957 edition of the *AOU Checklist of North American Birds*.

Having no subspecies, its geographically restricted breeding populations (Alaska and Siberia) are separated only by the Bering Sea. Long-billed Dowitcher breeds and winters farther north than its polytypic cousin, Short-billed Dowitcher. Long-billed also migrates earlier in spring and later in the fall, and it shows a marked preference for freshwater environments, in contrast to Short-billed's taste for tidal estuaries and mudflats. Long-billed also does not molt completely during migration like Short-billed, but instead molt is delayed until a wintering area is reached before some wing coverts and flight feathers are replaced.

Still the two dowitchers are very similar, especially in nonbreeding plumage, but Long-billed is easily distinguished by its call notes, a sharp *keek*, versus a mellow, 3-note down-slurred *tlu-tu-tu* for Short-billed. Long-billed usually gives its strong *keek* note when flushed, often repeated several times, and they frequently twitter to each other while feeding, unlike Short-billed, which are mostly silent feeders.

Spring migration is early February to late May, and birds migrate both during the day and at night. Peak migration through Texas is mid to late April, and through Western locations between late April and early May. Long-billeds arrive unpaired on breeding grounds in mid to late May where they engage in spirited aerial displays and chases, often involving multiple suitors. If shorebirds truly have songs, dowitchers sing them, with extended twittering notes strung together in musical fashion during

This species is more rotund, with a longer, more uniformly straight bill than the similar Short-billed Dowitcher, but it was not until 1950 that the two species were separated. This breeding female Long-billed Dowitcher from Salt Lake City in May shows a very long, straight, dark bill, brick red underparts, and mostly black upperpart feathers with narrow rust edges, unlike Short-billed Dowitchers' orange underparts and wider, orangish upperpart feather fringes.

△ Long-billed Dowitchers have deeper, more egg-shaped undercarriages and longer rear bodies than Short-billed. Note the clean white lesser underwing coverts which are streaked in Short-billed on the lower nonbreeding flying bird. A sleeping group of molting Long-billeds in March in e. Texas contrast with a single Short-billed Dowitcher's paler underparts that show spots rather than bars and a flatter back angle (second from left). Short-billed molts into breeding plumage later in spring than Long-billed.

their aerial displays. Parts of these songs are also heard during late stages of migration while birds are still on the ground.

Long-billed breeds primarily in high Arctic coastal tundra from ne. Siberia across the Bering Straits to w. Alaska and the North Slope, and east along the Beaufort Sea coast to w. Mackenzie Delta in the Northwest Territories in Canada. Nest locations are typically in wet grassy or sedge meadows near extensive flats and freshwater ponds.

The final pairing is monogamous, and pairs construct a cup-shaped depression lined with grass and moss at the base of a tussock in moist sedge tundra, often near open water. Here the female deposits 4 eggs that are incubated by both partners for 21–22 days. The nests are fairly deep and often damp at the bottom. Chicks leave the nest soon after the last egg hatches and are cared for by the male, who leads them to wetter areas offering food and more cover. Males may leave before chicks fledge (in 20–30 days).

◁ A Long-billed Dowitcher with short bill and fresh breeding plumage shares a freshwater marsh with Lesser Yellowlegs in High Island, Texas in April, where both are common spring migrants. Note the heavy barring with strong white fringes on the upper flanks of this Long-billed, which are lacking in breeding Short-billed Dowitchers.

A male Long-billed Dowitcher with a very short, straight bill walks on the Arctic coastal tundra in Kuparuk, Alaska in early July. Breeding in a small area of near-coastal Arctic tundra in Alaska and nearby Canadian Provinces, this bird's thin, rust upperpart feather fringes and heavy bars on the upper flanks have already worn off. A breeding male in flight from Alaska shows mostly black upperpart feathers with narrow rust fringes, a tail with wider black bars and narrower white and rust ones, and the classic white slash on the back.

When Kevin worked as a shorebird biologist on the North Slope of Alaska, Long-billed Dowitcher nests were the most difficult to find, with the nonattending member of the pair acting as a decoy. The nests were usually found only by rope dragging, as the incubating adult would sit tight on the nest without flushing, even with humans nearby.

During Kevin's 4 years as a shorebird biologist on Alaska's North Slope, the nests of Long-billed Dowitcher were the hardest to find as birds would sit tight on mostly hidden nests, even when biologists walked very close by. This bird never flushed from its nest even as this picture was taken at close range.

⬛ A composite photo shows a fresh juvenile male Long-billed Dowitcher (*left*) from Alaska in September and a mostly nonbreeding 1st-winter female in Texas in November. Note several dark retained juvenile feathers with rust fringes on the back of the 1st-winter female, which has a very long bill, well outside the range of any Short-billed Dowitcher.

◀ Long-billed Dowitchers are typically found in small to very large flocks away from breeding grounds. This group from Titusville, Florida in January shows the cohesive nature of these flocks, whose tight-knit group serves to deter aerial predators that are not able to focus on a single bird in flight.

▶ This great comparison of the two dowitcher species by Julian Hough shows the rounder body shape with distended, egg-shaped undercarriage and more front-heavy body structure with thicker neck of Long-billed Dowitcher (*left*) compared with a very similar plumaged *L. g. hendersoni* Short-billed Dowitcher in late August. When plumage details don't help with the ID of dowitchers, physical features often provide the identity of these similar species.

Fall migration is protracted, with adult females and failed nesters departing in July and adult males several weeks later. In northern Alaska, juveniles leave natal areas and flock to staging areas along the Beaufort Sea, where they feast mostly on fly and other insect larvae. Juveniles depart in September–October. Winter range is large, extending from central California south through much of Mexico to coastal Guatemala and El Salvador, and across the lower portion of the United States to Florida and North Carolina, but not south to South America like Short-billed. Along the Atlantic Seaboard, the birds are an uncommon migrant, and rare in spring.

Feeding in mostly smaller flocks than Short-billed, the birds probe muddy substrates with standing water in search of worms and other invertebrates. They may forage up to their bellies and sometimes will swim. When probing, birds may immerse their heads. In lesser used tidal areas, crustaceans, mollusks, and marine worms round out the diet.

Based on limited Arctic PRISM surveys in Alaska, where a majority of the population reside, along with various southbound migration counts in the Prairie Pothole region of the western United States, the total North American population of Long-billed Dowitcher was readjusted back to the previous 2001 estimate of 500,000 birds (Morrison et al. 2001). Because migration counts are highly variable from year to year, population trends are unknown.

▶ American Woodcock
Sclopax minor

BIOMETRICS 10–12½ inches long; wingspan: 16½–20¼ inches; weight: 135–211 grams (4.8–7.5 ounces)
STRUCTURE Fat, rounded body with large head; short, thick neck; long, straight bill; short legs
STATUS Common and widespread, but secretive and seldom seen, except when flushed or during twilight display flights. During heavy snow events, often seen feeding along sides of roadways.

- -

The early bird catches the worm. To no species does this adage apply more than to this plumpish, straw-billed shorebird. Long before American Robins are striding across suburban lawns, American Woodcock are probing sun-softened spring forest floors in search of their annelid prey, who are themselves feeling the procreative stirrings of spring (see page 21, lower photo). From mid-February to mid-March, nature centers all across the ne. United States and s. Canada kick off the field trip season with the near obligatory twilight "Dances with Woodcocks" field trip.

The pageant unfolds at a forest clearing as human participants shiver in the dusk-chilled air and listen for the rhythmic incantation that picks winter's lock upon the world. *Peent…peent…peent…*then up, up, up, up, up. It's Pan playing a reedy flute followed by a triumphant

▶ American Woodcock is one of only two shorebirds that are legal game. After seeing this soulful-eyed puffball, one is brought to wonder how anyone could hunt this bird. NJ, FEBRUARY

American Woodcock uses its long bill to probe deeply into the soil to locate its favorite food by using a flexible, sensitive bill tip to locate and grab earthworms. Meanwhile its large, laterally placed, binocular vision eyes scan the landscape for potential danger, even from behind. These secretive birds typically don't flush until they are underfoot, at which time they launch vertically into the sky.

NJ, FEBRUARY

distancing themselves from siblings while remaining generally close to the natal area until fall migration.

American Woodcock occurs only in Eastern North America, breeding in young (regenerating) forests generally east of the Great Plains, north to s. Canada and south almost to the Gulf Coast. They migrate in fall to regions south of the snowpack and the freeze line so that birds can continue to probe for earthworms. These plumpish, upland shorebirds shun shorelines and wetter habitats unless freezing temperatures force them to find water-softened soil they can penetrate with their long, tweezer-tipped bills.

Following a heavy snowfall in Cape May, NJ, where many birds attempt to winter, a woodcock was seen probing the snow-free earth beneath a dripping outside water spigot. During the winter of 2000, many woodcock perished when a heavy snow was followed by an ice storm, and they could not access any open ground to feed. Years of mild winters enabled them to overwinter successfully on Cape May Point, but the onset of severe weather in 2000 was too much for them to survive.

In migration and in winter, woodcock favors young poplar/aspen thickets but will use almost any moist forest environment. It is a shy and furtive species, and unless detected by a pointing dog, the superbly camouflaged birds sit tight and flush only when nearly underfoot. Their escape involves a near vertical climb, avoiding limbs until emerging from the canopy.

In winter during bouts of snow cover, woodcock will emerge from their usual forest haunts and feed along roadsides where the ground thaws out first, allowing for great views from a car. The eyes of woodcock are laterally placed to afford binocular vision in front and behind when their bills are buried in the soil (see photo left). In addition to earthworms, the birds consume snails, centipedes and millipedes, ants, and beetles.

This is one of two unfortunate shorebirds still considered gamebirds in the United States (Wilson's Snipe is the other) and legal to be shot for "sport." Annual harvest for American Woodcock in the United States and Canada averaged 314,634 birds between 2006 and 2010 (Cooper and Parker 2011). The federally established limit is three birds per hunter per day.

With no new information, population estimates of 3.5 million, with a range of 3–4 million birds, is retained (Morrison et al. 2006). Though declining over the long term, possibly due to the gradual maturation of eastern forests, regional populations have stabilized during 2001–2011, as measured in the Singing Ground Survey (Cooper and Parker 2011). Therefore, the current population is considered stable.

cascade of notes as hormonally fueled male woodcocks spiral back to their launch points. Displaying males may spiral upwards on twittering wings as high as 200–300 feet, and as the bird zigzags back to the ground, the outer flight feathers (primaries) produce three to six sets of descending twitters.

In larger fields (or singing grounds), multiple males often vie for a female's attention. In this polygamous mating system, females may visit multiple singing grounds and mate with multiple males, who share none of the obligations of parenting. Alone and as early as January in more southern locations, the female shapes a shallow depression in the dry leaf litter, usually close to a tree or shrub, and near the edge of a young upland forest. The 1–5 eggs are given no concealment beyond the incubating female's superbly camouflaged, leaf-patterned plumage.

If the eggs survive predation and desertion (as sometimes occurs in the case of premature nesting efforts), the precocial young hatch in 20–21 days, grow rapidly, and are nearly full-sized in 30 days. Since the probe-feeding technique used by woodcock is so specialized, young are dependent on the female for food for the first week, although juveniles have been seen probing in as few as 3–4 days and are fully independent in 31–38 days. At this time, they begin to roost at night in nearby fields, gradually

▲ An American Woodcock spreads its wings and raises its tail in a warning to nearby woodcocks not to come near its chosen feeding area. The stress of a snowstorm followed by an ice storm in Cape May, NJ in January 2000 resulted in many birds dying as they could not access open ground to feed.

▶ Thank goodness for open ground! This American Woodcock found an open grassy area to probe for food, which probably saved its life after a bad snowstorm blanketed Cape May, NJ in January.

▼ With superbly camouflaged plumage that matches the color of winter leaves, and black feathers on the back to approximate shadows, American Woodcock is virtually invisible when tucked into wood edges or thickets. This bird from s. NJ in February is watching Kevin without moving or even blinking an eye.

▶ Wilson's Snipe
Gallinago delicata

BIOMETRICS 10–11½ inches long; wingspan: 17¼–19½ inches; weight: 79–146 grams (2.8–5.2 ounces)
STRUCTURE Fairly chunky, chesty body shape with tapered rear body; shorter wings, tail, and legs than dowitchers; often crouches close to the ground
STATUS Common and widespread, but secretive and often hard to see

---

The well-known term "snipe hunt" does not reflect an actual group hunt of this common shorebird but refers to a practical joke that was popular in the 20th century, especially at summer camps and on *The Honeymooners* television show with Jackie Gleason. It involved taking someone out to a marsh or woods at night with a bag in their hand, along with a promise that if they stayed still and did not move, a Snipe would fly into the bag. This folly was just an excuse to trick someone into going out in the dark and standing still for long periods, and not a good way to spot a Snipe at all.

Snipes are chesty, pudgy, short-legged shorebirds that spend much of their lives in wet habitats, including marshes, lake edges, and pond borders. They move in and out of tall grass as they forage and stick close to cover to allow escape from danger. If they sense a possible threat, they sneak into dense cover, where they remain until all perceived danger has passed. Sleeping for much of the day, snipes are typically most active at dawn and dusk away from breeding areas.

This medium-sized, straw-billed marsh bird ranks as one of the most common and widespread of all shorebirds. It is found in appropriate habitat everywhere in North, Central, and n. South America at some point in the year, and even in some habitats that are not typical, especially during migration. Wilson's is also considered by some to be conspecific with the Eurasian Common Snipe *Gallinago gallinago*. It was Alexander Wilson himself who pointed out that the snipe that bears his name has 8 pairs of tail feathers, as compared with the Eurasian Snipe's 7 pairs, although they appear very similar in the field.

Spring migration takes place between late February and late May, with departure from southern wintering areas between late February and April. Wilson's breeds across boreal and subarctic regions from the Aleutian Islands to Newfoundland, and south across northern portions of the United States. Arrival at breeding sites varies geographically, with southern breeding sites being occupied by early March, but Arctic regions not until late May.

The ethereal "winnowing" sound that birds incorporate in their territorial display is produced by air flowing over

Wilson's Snipe are secretive birds that frequent densely vegetated marsh edges and other wet habitats, where they probe the substrate with long, straight bills. Besides American Woodcock, Wilson's is the only other shorebird allowed to be hunted. This bird feeding in wet grass near Lake Kissimmee, Florida in February is how snipe are rarely seen—in the open. Note the very long toes that allow it to traverse vegetation in wet locations.

▶ Migrating and tired Wilson's Snipe may set down in almost any open area, including lawns, highway medians, and roads where their camouflaged pattern does not avail them. This bird appeared on Kevin's lawn near Cape May, NJ on March 30 during migration and stayed only a few minutes due to a lack of vegetative cover.

▼ Not typically a flocking species, this group of Wilson's Snipe was flushed from a productive vegetated marsh edge in the Rio Grande Valley of Texas in November. After one bird flushes while uttering a gritty warning call, other birds take to flight. They then fly swiftly in a zigzag fashion, making them notoriously hard to shoot. With danger past, this group calmly flew back to the marsh.

the spread outer tail feathers, which vibrate when birds reach airspeeds approaching and exceeding 25 miles per hour. Winnowing flight begins as soon as birds arrive on breeding grounds in April or May, with males arriving 10 days to 2 weeks before females, which also winnow. The sound is a slow, repetitive hu...hu...hu...hu...hu...hu, sometimes likened to the call of Boreal Owl or the whir of waterfowl wings. This modulation is caused by wing-beats, which prompted some to initially conclude that the sound itself emanated from the wings. Wilson's most recognized sound, however, is the harsh, scratchy call made by the birds when flushed from nearly underfoot.

Breeding in forest bogs and marshy tundra, females construct several scrapes close to or surrounded by water and weave a lining of coarse grasses to build a nest up to 7 inches across and 3 inches deep. The nest is often placed atop or on the edge of a hummock and well hidden

by sedges, grass, or sphagnum moss. The female then lays 4 fairly large eggs, which she alone incubates, and adds more grasses to the nest between eggs. Upon hatching, the male typically leaves the nest with the first two hatch-lings, leaving the female to tend to the remaining two. Both adults go their separate ways, ending the monogamous pairing, and the nest mates also go their separate ways while following the adults. Unlike most shorebirds, the adults feed chicks until they are about 10 days old and able to feed themselves when their bills lengthen.

Fall migration takes place on a broad front from mid-July to December. Adult departure dates from breeding sites vary geographically. The first migrants in July are probably failed breeders, with peak migratory movements from September to mid-October in the West, and from mid-October to late November in much of the e. United States.

When on breeding grounds, as with this Wilson's Snipe from British Columbia in June, males seek out an elevated perch from which to launch their winnowing flight displays, or just to survey their territory. Wilson's breeds in forest bogs and marshy tundra from the Aleutian Islands to Newfoundland, and across northern portions of the lower 48 United States. Note the insect larva in the bill of this bird, which it is probably going to feed to its nearby young.

Superbly camouflaged to blend in with the wet grass habitats that snipes favor, Wilson's trust their cryptic plumage to conceal them. Birds sit tight when approached, then … *Yrrrch!* When flushed, Wilson's Snipe explodes into a low, zigzagging flight before leveling out, which distinguishes them from all other sandpipers. Then they drop rapidly into the marsh or occasionally circle back to land near their launch point. Despite their pudgy, unbalanced appearance, Wilson's is a strong, fast flyer and can approach speeds up to 60 miles per hour. While not typically a flocking species, multiple birds may occur in the same small patch of wet grassy cover, from which they flush individually.

Birds retreat south of the freeze line in winter, thus vacating most of their breeding range. Some birds, however, linger in areas where even a patch of unfrozen water allows their bills to probe in the wet, rich organic soil the birds favor. Birds use sensory receptors in the flexible tips of their long, straight bills to locate food under the surface of the substrate.

Wilson's forages in standing freshwater or the damp edges of bogs, swamps, swales, sloughs, and river edges. Wintering from s. Alaska and Massachusetts south to n. South America, birds occur across much of the lower 48 United States in winter, with the greatest concentrations

This is Wilson's Snipe as you typically see them—or don't see them! Sitting totally still until danger passes is a game snipes play to win. This bird with duckweed on its bill enjoyed the ample winter food at Wakodahatchee Wetlands in Delray Beach, Florida, but it would be migrating north soon after this March 4 photo.

Even on an open mud border, the camouflaged, earth-toned plumage of Wilson's Snipe blends in with the dark mud as it probes deeply under water to locate and secure aquatic invertebrates with its flexible, prehensile bill tip.

FLORIDA, JANUARY

of birds found in Louisiana, where snipes favor marshy pastures, rice fields, and flooded grasslands.

Mostly crepuscular (most active at dawn and dusk), birds feed by probing, often multiple times in the same place, in search of insect larvae and earthworms. With so much time dedicated to having their faces in the mud, snipe have evolved laterally placed eyes, giving them binocular vision fore and aft to help in the detection of predators, including humans. Wilson's Snipe is one of the two shorebirds that remain on the sport hunting game list. The other is American Woodcock.

The daily bag limit is 8 birds per hunter per day, and in 2018, 83,600 snipe were taken in the United States, with Michigan and Louisiana leading in harvest totals with 4,800 and 800 birds taken, respectively. Before the institution of bag limits, daily totals in excess of 100 birds per shooter per day were not uncommon, and the population suffered accordingly.

Arthur Cleveland Bent, writing in 1927, observed that "probably more snipe have been killed by sportsmen than any other game bird." He goes on to give the account of a seasoned Louisiana snipe hunter, James J. Pringle, a gentleman sportsman, not a market gunner, who killed 69,087 Wilson's Snipe between 1867 and 1887, with a single day record of 366 snipe shot on December 11, 1877. These figures attest not only to Mr. Pringle's skill, but also to the one-time abundance of this species.

The population is estimated to number 2 million individuals and is considered stable (Andres et al. 2012).

▶ Spotted Sandpiper
Actitris macularia

BIOMETRICS   7¼–8 inches long; wingspan: 14¾–16 inches; weight: 43–50 grams (1.5–1.7 ounces)
STRUCTURE   Short legs and bill, similar to "peep," but longer neck and tail; horizontal posture
STATUS   Common throughout interior North America and along coasts

It's the shorebird next door, although you may not necessarily be aware of this. This most widespread but surreptitious of shorebirds in North America is very likely incubating a clutch of eggs close to a fresh water source near you. Found from the Arctic to the southern states, this dapper, refreshingly well-marked sandpiper is easily distinguished with or without spots by its signature bobbing or teetering motion, which involves the entire body.

Like a nervous tic, the birds bob as they walk, most typically along the edge of any water body (lakes, ponds, beach wrack lines, and streams, including arid locations and urban parks). They also occur in man-made habitats, such as concrete drainage ditches, sewage ponds, and seawalls. The rate of teetering increases when birds are nervous, but stops when they are alarmed, courting, or aggressive.

Another interesting behavior of this handsome shorebird is the stiff, sputtering, staccato wingbeats interspersed with stiff wing glides that the birds exhibit when flying low from point to point. There is no explanation or

Breeding widely across North America, this refreshingly distinctive shorebird nests in open terrain or under vegetative cover bordering freshwater lakes and ponds. This well-marked Spotted Sandpiper walks on a downed tree near its nest in British Columbia in late June.

The stiff-winged, staccato flights of Spotted Sandpiper are distinctive and unique among shorebirds, and they are performed only in short sorties at breeding, nonbreeding, and migratory locations. During extended migration, Spotted's flight is strong and direct, which they need to carry them on nonstop, transoceanic flights to South America. BRITISH COLUMBIA, JUNE

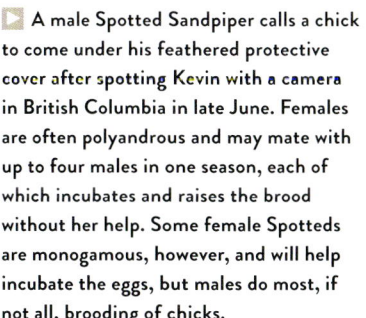

A male Spotted Sandpiper calls a chick to come under his feathered protective cover after spotting Kevin with a camera in British Columbia in late June. Females are often polyandrous and may mate with up to four males in one season, each of which incubates and raises the brood without her help. Some female Spotteds are monogamous, however, and will help incubate the eggs, but males do most, if not all, brooding of chicks.

function for this idiosyncratic wingbeat pattern, even though all members of this species perform it. In migration, the birds fly straight ahead with strong, steady wingbeats that don't resemble the staccato flight at all.

Spring migrants depart s. South American wintering locations between late February and late April and pass through n. South America from late April to late May. Arrival at northernmost breeding areas occurs in early June. Other breeding locations are accessed from late April to late May.

Female Spotted Sandpipers sometimes practice a breeding strategy called polyandry, where a female mates with up to four different males, each of which raises the brood without her help. The female is also the dominant member of the pair bond, with her testosterone levels rising up to sevenfold during the breeding season, which may explain her assertive behavior and sexual role reversal.

Unlike most shorebirds, females arrive first on the breeding grounds to establish and defend their territories. They can also store sperm for up to a month, so a male from a previous mating may father an egg laid by a female for a different male's nest. Not all female Spotteds practice polyandry, however, with some monogamous females pitching in to help with the incubation of eggs, but the female does little to help with caring for the chicks.

Spotted Sandpipers reach their highest breeding density on islands and other locations that offer protection from mammalian predators. There, in a scraped nest lined with dead grass near or under vegetative cover and typically within 100 yards of the water's edge, the female deposits 3–5 eggs. In a monogamous relationship, both males and females incubate the eggs for 21 days, but in a polyandrous one, only the male performs nest duties.

Chicks are hardy and leave the nest under the care of the male within hours of hatching, where they peck and scratch for food. They also start teetering very soon after emerging from the egg. Males attend the chicks for about 4 weeks, at which time they may be observed following adults along the edges of water bodies. Fledging occurs at about 15 days, with weak flight possible at this time.

Spotted is a solitary hunter and does not join feeding flocks. Prey include almost any creature small enough to be captured, but the bulk of their diet consists of adult and larval insects. Food is sometimes dipped in water before consumption. Most prey is visually located at the water's edge by walking or wading slowly, then rushing forward to secure sighted prey. Birds occasionally leap to catch flying insects, and they sometimes run with heads lowered along shorelines. Birds also feed on dry ground away from the water's edge and habitually perch on raised objects

A Spotted Sandpiper walking on a submerged log is molting to nonbreeding plumage on its breeding grounds in British Columbia in July. As a solitary species, the word "flock" does not even exist as a concept in Spotted Sandpiper's mind, and they don't join other shorebirds while foraging, either.

Juvenile Spotted Sandpipers, like this bird from Cape May, NJ in August, lack spots on the underparts and sport strong black bars with buff fringes on upperparts and tertials. Distinctive teetering and tail bobbing are captured as blur movements in this photo. No water? Not a problem for Spotted Sandpipers, which may run to secure prey, but water is never far from this species.

Nonbreeding and 1st-winter Spotted Sandpipers lack spots, but note the brown cowl around the neck and white slash near the wing bend. A 1st-winter bird (*left*) from Texas in November still shows juvenile-plumaged dark bars with buff fringes on its upperparts, while the nonbreeding bird from Andros Island in January has a tail that extends way past its wingtips, inconsistent with Spotted's typical shorter tail, and a feature used to help identify the closely related Eurasian Common Sandpiper.

(like fence posts, docks, or rocks). In tidal areas, birds may sit out high tides on rock jetties or snow fence posts.

Spotted Sandpipers that spend the winter in North America occur mainly along the coasts, and in tropical areas on riverbanks, sandy beaches, mudflats, mangrove swamps, and assorted wet forest habitats. In fall migration, beginning primarily with females in early July, birds move in a broad front across the continent and may use any convenient water source as a stopover area. Peak migration for adults is from mid-July to mid-August, and through the South in August, with juveniles moving about 3–4 weeks later. Some birds make nonstop transoceanic flights to South America, with first arrivals occurring in early July.

The term "flock" does not appear even as a concept in the mind of Spotted Sandpiper, but roosts of up to 1,000 birds have been documented in South America, where most juvenile Spotted Sandpipers spend the winter. Anyone who has watched the sputtery, stiff-winged local flight of Spotted Sandpiper should be amazed by the bird's capacity to reach Venezuela. Except during migration, most flights by this species fall short of 100 yards.

Given Spotted's widespread breeding range (largest of any North American shorebird), the absence of concentrated staging areas, and its secretive breeding behavior, the population estimate of 150,000 birds (Morrison et al. 2006) seems very conservative. Considering the estimated breeding population in Ontario, Canada is in the hundreds of thousands (Ross et al. 2003 in Andres et al. 2012), and extrapolation of Breeding Bird Surveys using PIF physiographic strata indicated the population might be 656,000 (Morrison et al. 2001), a population estimate of 660,000 is suggested, pending further study (Andres et al. 2012). Combined data sources indicate a stable population trend.

▶ Solitary Sandpiper
Tringa solitaria

BIOMETRICS 7½–9¼ inches long; wingspan: 22–23½ inches; weight: 31–65 grams (1.1–2.3 ounces)
STRUCTURE Chesty with tapered rear body shape; short to medium tapered bill; more compact than Lesser Yellowlegs with shorter legs, wings, and neck
STATUS Fairly common throughout range, but never in large groups

This dapper sandpiper with distinctive plumage, black-and-white tail, and bold eye-ring is a refreshing difference compared with the many look-alike sandpipers. As its name suggests, this slender, attenuated sandpiper with long, pointed wings is usually seen alone, or in the loose company of a few of its kind. Feeding in small, secluded pools, Solitary lives a mostly quiet life, shunning flocks or other gatherings of shorebirds.

Solitary was first described by Alexander Wilson in 1813, but it was 90 years before this enigmatic, iconoclastic shorebird's nest was found and the secret of its nesting habits was revealed. Wilson surmised this boreal breeder nested on the ground like other shorebirds, but he was proven wrong when, in 1904, a central Alberta homesteader saw a bird land in a low tamarack at the edge of a bog and enter a nest. Upon investigation, he found an old American Robin's nest containing 4 eggs that were not robin's egg blue in color, but white with reddish-brown blotches.

The eggs were collected, but it wasn't until the following year that the identity of the bird was confirmed (Bent 1962). That this dark-backed, puddle-loving sandpiper uses the old nests of songbirds to hold its eggs really should not have come as a surprise, as Solitary Sandpiper's Eurasian sister species, Green Sandpiper *Tringa ochropus*, also nests in this manner.

Of the world's roughly 85 sandpiper species, only Solitary and Green sandpipers routinely lay their eggs in abandoned songbird nests, particularly the nests of American Robin, Rusty Blackbird, Eastern Kingbird, Canada Jay, and Cedar Waxwing. Green Sandpiper also shares Solitary Sandpiper's penchant for feeding alone or in small groups, as well as an affinity for freshwater habitats that are too small or confining to attract most shorebirds.

Solitary Sandpiper, a freshwater obligate, may spend an entire day foraging in a puddle the size of a bathtub. Favored habitats include bogs, sloughs, roadside and drainage ditches, water-filled tire ruts, sewage pools, swamps, wet meadows, ephemeral ponds, and pond edges. It sometimes forages in the company of Least Sandpiper, another species that excels at feeding in marginal habitats shunned by other sandpipers. Solitary Sandpiper avoids seacoasts and is rarely found in tidal marshes.

Spring migration takes place between late March and late May. Birds begin leaving winter territories in se. Texas, the Caribbean, Mexico, and Central and much of South America in March, and they occur as individuals or in small groups across much of the United States in April (southern

◀ The aptly named Solitary Sandpiper typically forages alone and is virtually never found among shorebird flocks, unlike its *Tringa* relative Lesser Yellowlegs. A freshwater obligate, Solitary is rarely found in brackish or saltwater habitats and prefers to forage on the edges of small, quiet pools. NJ, MAY

Demonstrating its balance and agility, our smallest Tringa Solitary Sandpiper appears as acclimated to woodland habitats as many passerine species, even though its perching rear toe, now nonfunctional after eons of evolution, cannot help with holding onto branches. Occasional wing flaps assist with steadiness on small branches. Note the heavily patterned underwing coverts. ALASKA, EARLY JULY

This rare photo of a Solitary Sandpiper incubating eggs in an old American Robin's nest near Fairbanks, Alaska in early June shows this shorebird's unusual and unique proclivity for using old songbird nests in boreal forests to lay their eggs. It is presumed that precocial downy chicks jump from nests in Wood Duck fashion, even from nests up to 40 feet above the ground! Thanks again to Ted Swem for this image taken near his woodland home.

tier states) and early May (northern states), with birds reaching breeding grounds mostly in mid to late May. Some one-year-old birds remain on the wintering grounds through the summer, while most apparently migrate north to the breeding grounds. Breeding range extends in the boreal-forest belt from Alaska east to New Brunswick, where birds establish territories in spruce forests with muskeg bogs, and where they use old songbird nests up to 40 feet high.

Solitary Sandpipers may spend hours foraging in roadside puddles the size of bathtubs, but this breeding bird is feeding in a quiet pond in West Paterson, NJ in May. Birds are typically seen alone or in small groups, but Kevin once witnessed a group of 11 migratory birds in the Dry Tortugas, Florida in April after a storm.

With breeding territories as large as 100 acres, male Solitary Sandpipers aggressively defend nest areas, with any intrusions usually resolved with threat displays like this one. In early July, however, this display might be an animated protest over an intruder near its hatched young. In the event of a true fight, extensive pecking occurs until one bird gives up. FAIRBANKS, ALASKA

△ The essence of a Solitary Sandpiper's arboreal habits on the breeding ground is captured perfectly in this moody photo where the bird looks totally at ease on a small, broken tree limb. On the late date of July 4 in Fairbanks, Alaska, the precipitation falling is rain, not snow.

The female may refurbish old nests or use them as found. Males establish territories as soon as they arrive in spring, some as large as 100+ acres, and defense of these territories can be quite aggressive. Conflicts between males over territories are often resolved through threat displays; fights involve extensive pecking until one gives up. Males display over the territory by rising slowly into the air on quivering wings, spreading the tail and singing, and then slowly returning to the same spot. Both parents incubate the 4 eggs, and it is presumed that the precocial and downy young jump from the nest in Wood Duck fashion, although precious little is known about their growth and development.

Fall migration takes place through the US interior or offshore, with no significant coastal concentrations.

▷ White spectacles impart to Solitary Sandpiper the suggestion of a stare. This juvenile may be perplexed by the absence of water that was here only yesterday, but small ephemeral freshwater pools are Solitary's habitat of choice. Juvenile plumage differs from adults by its uniform brown wash on the head and breast sides rather than defined dark streaks, and by smaller off-white to buff spots on the back. AUGUST, NJ

▲ Two nonbreeding Solitary Sandpipers from the Rio Grande Valley of Texas in November show blurry, diffused streaking on the breast and face and smaller, more reduced spotting on the upperparts. A juvenile at right shows strong buff spotting on back, which may suggest the Western subspecies T. s. cinnamomea from Bolivar, Texas in April, as both subspecies occur together on wintering grounds.

Adults begin to depart breeding areas in late June, with peak passage through northern tier states between mid-July and mid-August and through the south in August. Failed breeders head south before successful adults. Southbound juveniles depart the breeding grounds in late July and early August. Migration takes place primarily east of the Rocky Mountains. Solitary migrates mostly at night, either singly or in small flocks. Some birds make transoceanic flights from New England to South America, while others take the trans-Gulf route.

In winter, Solitary also occupies muddy shorelines of lakes, mats of floating vegetation, marshes, rice fields, lake edges, and sandy beaches along rivers. Birds seem particularly drawn to turbid water, which they explore with a quivering foot, or while wading up to their bellies. Tame and confiding, Solitary often allows close approach, but if startled, birds flush vertically amid a flurry of high-pitched, rising *peet, wheat, wheat* notes, only to flit to the other side of the puddle, or to fly far over the forest to another choice feeding area. Its diet is mostly aquatic insects and their larvae but also includes small crustaceans, small mollusks, and frogs away from breeding areas.

Two subspecies are generally recognized (Moskoff 2011 in Andres et al. 2012), *T. s. solitaria* and *T. s. cinnamomea*, which occupy mostly eastern and western parts of the breeding range, respectively, but have been found together on wintering grounds. Morrison et al. (2006) argued that while there was little basis for assigning separate population estimates to the subspecies, their relative distributions suggested an approximate ratio of 2:1 *solitaria* to *cinnamomea*. The total population estimate for both subspecies together is 189,000, though all sources for population trends are highly variable among years (Andres et al. 2012).

▶ Wandering Tattler
Tringa incanus

BIOMETRICS 10½–12 inches long; wingspan: 20–22 inches; weight: 100–140 grams (3.5–5 ounces)
STRUCTURE Slender and attenuated with short legs and medium-length, straight, tapered bill; horizontal stance
STATUS Fairly common but strictly coastal; usually seen singly or in small groups; very rare inland

The implication of its name notwithstanding, this medium-sized "rockpiper" in a plain gray wrapper is seldom seen away from Alaskan, Yukon, and Siberian breeding areas, or Pacific and Hawaiian Archipelago coasts. While described in 1789 from a bird collected in the Society Islands, it was not until 1912 that its nest was discovered on the Arctic coast of the Yukon, and it remains one of our most enigmatic and understudied shorebirds.

The name "tattler" relates to the bird's ringing and repeated alarm calls that alert other birds to a hunter's presence, while the name "wandering" refers to this species' occurrence on islands across vast portions of the Pacific Ocean. Bobbing its tail as it hunts for food and calling as it flies, Wandering Tattler is known to most people as a denizen of rocky Pacific coasts, where in winter it joins Black Oystercatcher, Surfbird, Rock Sandpiper, and Black Turnstone on rocky headlands above the wave zone.

When foraging, it is commonly alone in its linear territory, searching for insects, small mollusks, and crabs in the intertidal zone. Tattlers are nimble, active feeders that either pick from the surface of rocks or probe into clusters of oysters, mussels, and algae. The birds cover ground quickly and may run to capture elusive prey items in

▶ A Wandering Tattler as you often don't see them (in breeding plumage), unless you visit Pacific coastlines in late spring or early summer. These perpetual-motion pipers seem to be ever moving while actively patrolling rocky shores. When paused, they bob or teeter their bodies in a distinctive manner. This breeding adult in Homer, Alaska in May is poised to transition from rocky coasts to the fast-flowing mountain stream habitats where they breed.

▼ Wandering Tattler is a member of the "rockpiper guild," forming an alliance with Black Turnstone and Surfbird (shown here) away from breeding grounds (California, July). With landing gear down, a breeding bird at right shows the very long wings that allow this species to make pan-Oceanic migratory flights of up to 7,500 miles to islands and land masses in Oceania, including Hawaii and the Galápagos Islands. Others winter along Pacific coastlines from California to Ecuador.

◀ An ungainly pose with well-nourished, pear-shaped body and undersized head shows the stealthy stalking behavior of Wandering Tattler on rocky shorelines. Trying to hide in plain sight, this breeding bird from Homer, Alaska in May is picking insects and invertebrate prey from the wave-washed rocks.

receding waves, or jump to catch flying insects. When alarmed, it often bobs its head and foreparts up and down like all members of the *Tringa* shorebird group, where it was recently placed.

While allied to the *Tringa* sandpipers, Wandering Tattler is most closely related to the Asian Gray-tailed Tattler, *Heteroscelus brevipes*, which it strongly resembles, and with which it is sometimes considered conspecific. In the Russian Far East, the breeding ranges of Wandering and Gray-tailed Tattler overlap, but with limited interbreeding.

Wandering Tattler is an intermediate- to long-distance migrant. Some birds migrate mostly through coastal areas, while others may cross up to 7,500 miles of open ocean.

Spring migrants depart southernmost American wintering areas in early March. Peak movements take place along the Pacific Coast of Mexico and s. California between early April and mid-May; through the central and n. Pacific Coast in early to mid-May; and through s. Alaska in mid to late May. Arrival at breeding areas is mostly mid to late May, but sometimes as late as mid-June.

Breeding locations include Alaska and Yukon Territory as well as Siberia, where birds nest in shrubby alpine tundra above the tree line near rocky or gravelly streams and lakes. Birds are widely dispersed in steep-sided valleys in the alpine zone. The nest is placed in a natural depression among rocks and lined with a few twigs, or constructions

▲ Seen in flocks only during migration, this group from Homer, Alaska in May is getting ready to move to its breeding grounds in rugged landscapes in Alaska, the Yukon, and the Russian Far East, where montane tundra is cut by glaciers and streams. The flight of single birds, while strong, is typically low and weaving.

◄ In winter, Wandering Tattler forages along rocky coastlines rich in marine invertebrates. This wave-flushed individual in California is heading for higher ground. Named for its high-pitched alarm call, the "wandering" part of its name comes from this bird's occurrence on islands across vast portions of the Pacific Ocean in winter.

A distinctive silver-gray juvenile plumage with pale upperpart feather fringes and gray mottled chest and flanks is shown on this Wandering Tattler from Half Moon Bay in California in November. Migration of juveniles is from mid-August to late November, and some young birds spend up to 2 years on wintering grounds, returning in their third year to breed.

that are more elaborate, involving interwoven leaves, roots, and twigs (more intricate than is typical with most *Tringa* sandpipers). Both parents incubate the 4 eggs for 23–25 days. Young leave the nest with both parents (until females depart) and are able to walk, feed, and "bob" within a day of hatching. Sustained flight is possible in about 21–25 days.

Fall migration takes place from early July to November, with mostly adult females beginning their southbound migration in early July. Wintering sites in Central and South America may not be reached until late September or October. Peak numbers occur from Alaska to California between late July and mid-August, and in the Hawaiian Archipelago between mid-July and late September. Migration of juveniles lasts between mid-August and late November.

Birds winter along the Pacific Coast from California to Chile, but some make a pan-Oceanic flight from breeding grounds to the islands and landmasses of Oceana (including Hawaii and the Galápagos Islands), a distance of up to 7,500 miles. Here it is one of the most typical shorebirds found on beaches, rock outcroppings, coral reefs, and river mouths as well as on man-made habitats such as jetties, piers, pilings, and bulkheads. In North America in winter, Wandering Tattler is mostly associated with rocky, high-energy intertidal habitats.

Juveniles migrating after adults may spend up to 2 years in nonbreeding wintering areas before returning to breed in their third year. The global population is estimated at 10,000–25,000 individuals, most which breed in North America (Andres et al. 2012).

Lesser Yellowlegs
Tringa flavipes

BIOMETRICS 9¼–10 inches long; wingspan: 23½–25½ inches; weight: 48–114 grams (1.7–4 ounces)
STRUCTURE Delicate with slim chest and smooth, gently rounded body contour; small head with straight, needle-like bill; long legs and neck
STATUS Common at interior and coastal sites in fall; mostly midcontinent in spring; short- and long-term declines evident

Whether in the air or wading through the shallows, everything about this elegant wader seems perfectly choreographed, an elegance that makes the movements of Greater Yellowlegs appear stiff and bumpkin-like. With a gently rounded body shape, evenly balanced weight distribution, and proportionally long neck, Lesser Yellowlegs is the picture of grace and style as it moves casually around a shallow pool. It is smaller with a shorter neck and has a smaller, more needle-like bill than Greater Yellowlegs, and it walks in a deliberate, high-stepping manner, occasionally darting forward in pursuit of prey.

This dainty "marshpiper" favors shallow, weedy, freshwater and brackish wetlands during migration and winter, including lake and pond edges, flooded pastures, marshes, low saline mudflats, wet meadows, sewage pools, and flooded agricultural fields, but it avoids saltwater locations such as beachfronts and tidal flats. A more

Smaller, slighter, and more deliberate feeders than Greater Yellowlegs, Lesser Yellowlegs winters along the s. Atlantic, Pacific, and Gulf coasts as well as Central and South America, the West Indies, and Caribbean Islands. This breeding adult was foraging in Texas in April shortly before Lessers depart for their taiga breeding grounds.

A great comparison of breeding Lesser and Greater Yellowlegs shows Lesser's more graceful, evenly rounded body shape compared with Greater's larger size; longer, thicker bill and legs; rangy, bottom-heavy body shape; and Adam's Apple look created by a retracted longer neck. Since plumage may be virtually identical in adults of both species, these different physical features are valuable when dealing with only one yellowlegs species. FLORIDA, LATE APRIL

deliberate feeder than Greater Yellowlegs, Lesser picks at prey with its needle-like bill, whereas Greater uses its longer, sabre-shaped bill to stab the depths. Less frequently, Lesser probes into mud or sweeps the bill back and forth through water, where it captures small fish. They often associate with dowitchers and Stilt Sandpiper when foraging because of similar habitat preferences.

More social than Greater Yellowlegs during the nonbreeding season, Lesser often travels in flocks of six or more but can number in the thousands at migratory stopover sites. Migration occurs across both interior and coastal regions, though always more numerous along the coast, particularly in fall. In typical foraging posture, Lesser leans forward with its bill held at a 20–30-degree angle to the substrate and will occasionally swim to feed on surface organisms. Groups usually forage in a loose, spread-out fashion, and they do not form tight groups to corral fish, as does Greater Yellowlegs.

Spring migration takes place between late February and late May, with northbound adults departing the wintering areas in late February and early March. Spring migrants are more common in the Southeast and continental interior,

In migration and winter, Greater and Lesser Yellowlegs (*right*) often feed with Stilt Sandpipers, whose longish legs accommodate the water depth favored by yellowlegs. This photo from early April in Texas shows the typical foraging style of Stilt Sandpipers with body tipped forward and tail upward, and bills inserted fully into the underwater substrate. Success! The Stilt at left has a small morsel in its bill.

Lesser Yellowlegs typically forages in a loose, spread-out fashion and does not form tight groups to corral fish like Greater Yellowlegs. Generally more social than Greater Yellowlegs, these juveniles from NYC in August are feeding during migration, as are the Short-billed Dowitchers behind. While yellowlegs pick for their food, dowitchers probe.

"I give up!" is what the lower Lesser Yellowlegs should be shouting to the upper bird. Lesser is an aggressive defender of its feeding territory during migration, even when there is endless similar habitat. Without arms to fight with, birds try to land on top of their opponent, which usually signals defeat for the lower bird, but sometimes pecking of the neck and head occurs if the conflict escalates. Adults in NYC in August.

and less common in the Northeast and Pacific Coast. Peak numbers reach the Gulf Coast by mid-April, and the northern tier states by early May. Spring migration of 1st-year Lessers averages later than adults, with some birds remaining on wintering grounds through their first summer, while others migrate partway or all the way to breeding grounds.

Lesser feeds both day and night and is a fierce defender of its small feeding space during migration. Birds will face off in threatening fashion before one jumps high in the air while trying to land on the back of the other, where it can peck at the neck and head. Usually the conflict ends without serious consequences when one of the combatants retreats.

Lesser Yellowlegs breeds from Alaska east to the James Bay region of Quebec, primarily in open boreal forest and forest/tundra transition habitats. Arrival on the breeding grounds takes place between late April and early June. Birds reach breeding grounds in small flocks, and monogamous pairs form within a few days.

Nest initiation is primarily from mid to late May in milder areas of Alaska, and early to mid-June in central Canada. Four eggs are typically deposited in a moss-lined depression on the ground or on a dry ridge, often near a log or stump. Both sexes incubate and tend to young, which fledge in 18–20 days. Adults often sit in the tops of spruce trees to survey their territory, and the male's animated flight display often begins and ends at a treetop perch, usually accompanied by singing throughout.

Fall migration takes place between mid-June and November, with most adult females departing breeding areas by early July and most adult males by mid-July. Failed breeders may start to head south by mid-June. Southbound juveniles depart the breeding grounds between late July and mid-August and head to both coastal and interior sites. Lesser winters mostly in coastal areas from central California and s. New Jersey south through much of South America, though it is a regular migrant through interior North America, where it can be numerous at favored sites. Prime staging areas in British Columbia and Saskatchewan may host more than 1,000 birds.

Calls of Lesser and Greater Yellowlegs differ substantially, with Lesser giving a soft, plaintive one-note call (*chew*), which may be given repeatedly when they are disturbed or alarmed, while Greater gives a loud, strong, three-note, descending call (***chu**–chu–chu*), or a single, emphatic note if alarmed on the ground. If Lesser is truly

◄ Lesser Yellowlegs often sit on top of small spruce trees in the taiga where boreal forest meets tundra. They survey their breeding territory from these perches, or they land on them after performing an extended courtship flight. This Lesser is calling before or after one of these elaborate flights, and vocalizing may continue for long periods. CHURCHILL, MANITOBA IN JUNE

▼ Fresh juvenile Lesser Yellowlegs have a smooth brownish wash to the upper breast that is lacking in Greater, along with a brownish back and wings with petite off-white spots. Long, angular wings allow Lesser to fly swiftly and make quick turns, which help when avoiding aerial predators. Long, yellowish-orange legs, feet that extend past the tail, and an evenly balanced body and white rump complete this flight profile. AUGUST, NYC

△ Odd bedfellows for sure! An unusual photo of a juvenile Lesser Yellowlegs, whose brownish upper breast wash has worn to blurry streaks, calmly foraging beside a crisp juvenile Red-necked Phalarope at the Jamaica Bay Wildlife Refuge's East Pond in August. Right photo shows a graceful juvenile Lesser with evenly rounded body contours as it feeds on floating mats of vegetation with the help of occasional wing flaps. NYC, AUGUST

alarmed on the ground or when taking flight, the calls can increase in both volume and urgency, but they are always the same pitch and even cadence. Both species often call when taking flight.

△ This Lesser Yellowlegs transitioning into breeding plumage in April in Texas appears dapper and serene as it rests on one leg after foraging in the nearby marsh. Always a picture of grace, they never seem to be messy or out of sorts with respect to plumage.

Despite their similarities, Greater and Lesser Yellowlegs are not closely related genetically in the way Long-billed and Short-billed Dowitcher are. While both are sandpipers in the genus *Tringa*, the larger, more robust Greater Yellowlegs is more closely related to Spotted Redshank, *T. erythropus*, whereas Lesser is more closely related to Willet. The similarities between the yellowlegs are based on convergent, not divergent, evolution, which means they evolved from separate ancestors and developed similar physical traits over time to adapt to similar environmental conditions.

Like many shorebirds, Lesser suffered from market gunning in the 19th and early 20th centuries. Edward Howe Forbush relates an account of 106 Lesser Yellowlegs killed with a single discharge of shot (Giraud 1844, in Forbush 1912). The birds recovered somewhat, and by 1947, Ludlow Griscom (n.d.) reported 4,000 birds on Monomoy Island, Massachusetts, on August 6; however, continued loss of their preferred wetland habitat has contributed to recent declines.

Recent population estimates that include Breeding Bird Surveys increased the global numbers of Lesser Yellowlegs from 400,000 to 660,000, which includes a 20% decline noted by Morrison et al. (2006); however, signals of short- and long-term declines are evident in a variety of datasets, and these sources indicate a significant decline going forward (Andres et al. 2012).

■ WILLET *Tringa semipalmata*

Willet is one of the most understudied birds in North America despite its accessible nesting and wintering locations. The two distinct subspecies have a number of differences regarding their nest locations and habitat preferences, migratory timing/strategy, molt timing, voice, and wintering locations (with some overlap), but their separate speciation is pending because of the lack of supporting viable research, DNA analysis, and published papers, and not for disputes relating to their distinctiveness.

Sexual dimorphism (differences between males and females) with respect to size, body shape, and bill structure is apparent in Western Willet but not typically in Eastern, with female Western Willet averaging about 20% larger in body and bill size compared with male Western and all Eastern Willets, and roughly the same percentage of differences with respect to leg length. Differences in appearance are moderately distinct in all plumages, but some individuals are problematic, especially nonbreeding birds.

▲ While male Western Willets often appear physically similar to Eastern Willet; female Westerns do not. Differences include up to 20–25% longer legs and bills on Western, with bills thinner and more tapered and pointed than Eastern's thicker, blunt-tipped bill. Western's football-shaped body, similar to Marbled Godwit, differs from Eastern's more compact, front-heavy body structure with foreshortened rear end, which results in a more upright stance in Eastern. HIGH ISLAND, TEXAS, MARCH

▼ This photo of Eastern and Western Willets together in a NJ salt marsh in late July shows the darker overall plumage of Eastern in midsummer and noticeable physical differences, including Western's longer legs, longer bill with thinner base that tapers to the tip, and longer, football-shaped body. An *L. g. hendersoni* Short-billed Dowitcher is at right.

▶ Eastern Willet
Tringa semipalmata semipalmata

BIOMETRICS   12½–14 inches long; wingspan: 21½–24½ inches; weight: 199–263 grams (7.1–9.3 ounces)
STRUCTURE   Stocky and compact compared with Greater Yellowlegs and Western Willet; bulky chest and foreshortened rear body compared with Western Willet's long, attenuated rear, and shorter, thicker neck and bill than Western
STATUS   Common and conspicuous breeder in Atlantic and Gulf Coast coastal marshes; population appears stable

Loud and boisterous on the breeding grounds is the playbook for this large, common shorebird with distinctively marked black and white wings. Visiting one of their breeding sites along the Atlantic or Gulf coasts is an experience not soon forgotten. After a day in Eastern Willet infested marshes, your ears are ringing and your thoughts regarding Willets are dark. While Western Willet (*T. s. inornata*) are just as boisterous, they don't occur with the same density at western interior freshwater locations, so the overall noise and frenetic activity is much reduced.

But there is little reticent about Eastern Willet. As soon as the heads of researchers come into view, nesting birds go aloft and fly straight at the trespassers, making repeated dives at unprotected heads. With young present, the displays are even more aggressive and frenzied, and birds from adjacent territories cross boundary lines to join in the fray. There may be no other North American shorebird that behaves so aggressively during the breeding season toward intruders.

The nominate Eastern Willet (*Tringa semipalmata semipalmata*) breeds in salt marshes and mangroves, and on barrier islands and beaches from Newfoundland to ne. Mexico, with isolated resident populations in the Bahamas, Greater Antilles, Cayman Brac, and the island of Los Roques off Venezuela. The birds were nearly extirpated along the Atlantic Seaboard because of market gunning and egg collection in the 19th and early 20th centuries. Edward Howe Forbush wrote in 1912 that Eastern Willet "formerly bred all along the Atlantic coast from Nova Scotia … to the Gulf of Mexico." John James Audubon wrote in 1871, "eggs afford excellent eating," and young "grow rapidly, becoming fat and juicy, and by the time they are able to fly, afford excellent food."

Fortunately, a vestigial population survived in Nova Scotia and along New Jersey's Delaware Bayshore, and the population recovered. The Migratory Bird Treaty Act of 1918 banned market hunting and marked the start of the Willet's comeback. Now greatly recovered, Willets are approaching the numbers seen by Alexander Wilson, who said of the bird: "Breeds in great numbers on shores of New York, New Jersey, Delaware and Maryland (Wilson 1818–1829).

Migrating mostly at night, Eastern Willet takes a primarily transoceanic route between South American

◀ A heavily marked breeding Eastern Willet from NJ in April shows the diagnostic stout, straight bill with pinkish base and heavy underpart markings that are noticeably darker and denser than breeding Western. This large shorebird is a common fixture on Atlantic and w. Gulf saltwater coastlines, but only during the spring and summer breeding season..

◀ Encroach on an Eastern Willet's breeding territory and the results are dramatic and vocal. Pairs from adjacent territories will often join in, crossing territorial boundaries to help drive off intruders. If they don't physically drive you away, their constant loud protest calls will. FLORIDA, APRIL

▽ Eastern Willet lays large pale green eggs with dark splotches, often in a shallow depression in marsh grass adjacent to tidal areas. A young juvenile at right in early September in NJ shows a short, extra-thick bill with blunt tip and dark scapulars that contrast with paler wing coverts. Upper breast streaking is sparse and fine, unlike the heavy markings on adult. This late date is very unusual for Eastern Willet fledglings.

▽ Rarely seen in flocks except during migration and postbreeding staging, a group of 30 Eastern Willets arrived at High Island, Texas on March 15 after flying from wintering areas in Brazil. Birds arrive still molting into breeding plumage and quickly disperse to establish breeding territories. Note the tiny Least Sandpipers in the foreground compared with this large *Tringa* sandpiper. Eastern Willet lays large pale green eggs with dark splotches, often in a shallow depression in marsh grass adjacent to tidal areas.

Eastern Willet flight displays at breeding sites are noisy and long, often lasting for hours on end, much to the dismay of researchers' ears. Flying high in place on stiff, shallow, quivering wings, the *peer-willet* call that earned this bird its name is repeated over and over and over again. TEXAS, APRIL

wintering locations and North American breeding grounds. Spring migration occurs mostly between early March and early June, with Pacific migrants arriving along the Gulf Coast from early to late March. Peak numbers depart South America in mid-April and arrive along the s. Atlantic Coast in mid to late April, and in New England in late April/early May, although migration continues steadily throughout May.

Eastern Willets arrive on territory with their beaks wide open and their boldly patterned wings in a frenzy. Incessant flight displays with quivering wings are accompanied by their very loud, namesake calls (*peer willet*, *peer willet*, *peer willet*), and these displays seem to go on with little pause all day long. If other Willets venture into the space of a territorial bird, a chase of high speed and agility ensues until the intruder is driven from the area. Were it not for such boisterous displays, these plain grayish-brown marsh birds might easily go unnoticed.

Eastern Willet feeds by day and night, with primary foods being crustaceans, particularly fiddler crabs and mole crabs, but other food sources include aquatic insects, marine worms, small mollusks, and fish. They rarely form flocks, except in early summer after breeding and prior to migrating, and during early spring flights to breeding

Strutting in a stately fashion with neck raised and tail fanned, a male Eastern Willet roams his beach nest territory with high steps while displaying for his mate. While not colonial nesters, multiple pairs of Eastern Willets may share prime beachfront nest habitats without much conflict. High breeding density occurs from High Island to s. Galveston Island and points south along the coast of e. Texas. TEXAS, APRIL

⚠ Nonbreeding Eastern Willets (*left*) are quite bland, with grayish overall plumage and whitish belly that are similar to Western Willet. Nonbreeding birds are virtually never seen in North America, with this bird part of a resident population on Cayman Brac in early March. Upon landing, Eastern Willet habitually raises its wings upward, similar to Upland Sandpiper. NJ, MAY

areas on the Eastern Seaboard, when up to 50 birds may occur in a single flock. Nest initiation ranges from mid-April in the south to late May–early June in the north. Incubation is 24–26 days, and both adults share parental duties, though females typically abandon young and males after 2 weeks. Young can fly after 28 days.

Fall migration is complex, with Atlantic Coast Eastern Willets heading south over the Atlantic in late July and early August. Winter range is poorly known, but it is apparently centered in e. South America, especially Brazil. Birds often use the Outer Banks of North Carolina as a staging area before continuing to coastal Brazil. The w. Gulf breeding Eastern Willets (w. Louisiana south to Mexico) migrate earlier, mostly in July, and head south along the Gulf Coast until Panama, after which they cross over the Isthmus before continuing to the west coast of South America as far as Peru. This suggests a possible subspeciation, or at least a distinct recognition of breeding populations among Eastern Willet.

Eastern Willet from Atlantic coastal areas winter mostly in the extensive mangroves and adjacent tidal flats and creeks of coastal Brazil, while Western Gulf breeders utilize a variety of coastal habitats in winter. There are few, if any, viable records of Eastern Willet in North America in winter.

Recent population estimates place Eastern Willet at 90,000 birds (Andres et al. 2012)

▶ Western Willet
Tringa semipalmata inornata

BIOMETRICS 13½–16½ inches long; wingspan: 23½–28½ inches; weight: 203–339 grams (7.2–12.1 ounces)
STRUCTURE Lanky and long-legged with a football-shaped body similar to Marbled Godwit; longer bill than Eastern, with thick base and tapered lower outer third, giving the appearance of a slightly upturned bill; longer neck and proportionally smaller head compared with Eastern Willet; mostly horizontal posture compared with Eastern's more upright one, and a longer, more attenuated rear body.
STATUS Common in many coastal areas during migration and winter; more sparsely distributed as breeder in western areas, and a less common migrant through the interior.

Unlike their closely related subspecies Eastern Willet, which breeds in brackish to saltwater coastal marsh habitats, Western Willet breeds in freshwater prairie habitats in the Western United States and Canada. Given the wide-open spaces of their breeding range, the aerial duels between competing pairs of birds—a trademark of Eastern Willet in coastal areas—is not typically a part of Western Willet's playbook.

They are, however, just as aggressive toward intruders into their breeding territories, as Kevin witnessed near Salt

While still considered conspecific with (the same species as) the smaller and more heavily marked Eastern Willet, Western Willets breed and winter in distinctly different habitats and locations, and they deserve separate species status in Kevin's opinion. This breeding adult Western Willet showing more lightly patterned underparts than Eastern with a buff wash to the flanks, a longer, thinner all-dark bill, and grayer overall plumage than Eastern's browner look was photographed on the Texas coast in April where they overwinter.

Lake City in May when a pair of Western Willets joined a few Long-billed Curlews in chasing an adult Golden Eagle from their airspace. With loud vocalizations that are only slightly different from Eastern Willet, these prairie breeders can raise quite a fuss when unwanted intruders appear.

Differences from Eastern Willet are pointed out at the beginning of this willet profile, but the most obvious one is the very large size with noticeably longer legs and bill in female Western Willet (up to 20% larger and 20–25% longer legs and bill). These birds walk and run with a distinctive loping stride, similar to a 7-foot-tall basketball player, and their overall appearance is lanky and elongated compared with Eastern's more compact and foreshortened body shape.

Some birds, mostly males, can be quite similar in size and bill shape to Eastern Willet, but their rear bodies are more elongated, and their overall shape is more football-like, similar to Marbled Godwit. Eastern Willet also usually shows a more overall stocky bill with a thicker base and thicker tip, and in breeding season the bill has a pinkish base versus an all-dark bill in breeding Western, and a pale grayish bill with dark tip in nonbreeding birds. Eastern also attains breeding plumage sooner than Western on average.

Spring migration occurs primarily between late March and mid-June, with peak numbers passing through the s. United States in late April and s. Canada in early May. Subadult birds often remain on the wintering grounds through their first summer, and a small percentage through

Standing tall on the Texas coast, a nonbreeding female Western Willet towers over a breeding Eastern Willet and transition Black-bellied Plovers but not the Black-necked Stilts standing behind in shallow waters near High Island, Texas in mid-March. All the physical differences covered in multiple captions are evident here. Western Willet attains breeding plumage later than Eastern on average.

▶ Two nonbreeding male Western Willets (*center*) show physical features and bills unusually similar to those of a breeding Eastern Willet at right. A pinkish bill base on Eastern differs from the pale gray bill base on Western, whose bills are slightly longer and more tapered toward the tip. This is about as physically similar as these two species will ever get. A Long-billed Dowitcher sneaks away at left in High Island, Texas on March 12.

▲ A fresh juvenile Western Willet (*left*) from NJ in August shows a godwit-like profile and neat, crisp, ginger-toned upperpart feathers with intricate internal markings, which differ markedly from juvenile Eastern's dark scapulars (see page 247, middle right photo). A nonbreeding Western from CA in November shows a bold black, white, and gray wing pattern that adds a dramatic touch to birds in flight.

their second summer. Migration takes place mostly at night, but also during the day in certain conditions.

Western Willet breeds in a broad range from s. Alberta and Manitoba to n. California and Colorado. They nest primarily in sparsely vegetated prairie wetlands or adjacent semiarid grasslands, with nest initiation in early May in the south of their range, and early June in the north. Nest site is typically in a sparsely vegetated area adjacent to a windbreak, such as a piece of wood, dried cattle dung, or a rock. During the breeding season, insects are the primary food source for Western Willet in the prairies.

The courtship display is a duet, with the male rising first and hovering over the territory on quivering wings.

He is then joined by the female, who hovers and descends to the ground with the male. Birds sometimes display in small groups with other pairs or unmated birds. Both adults share parental duties, though the female typically abandons young and male after 2 weeks, leaving the male to rear chicks by himself for the next 2 weeks. Incubation is 24–26 days, and young can fly after 28 days.

Fall migration occurs primarily between mid-June and late October, with peak numbers passing through the Pacific, Gulf, and s. Atlantic coasts between late July and mid-August. Western Willet migrates to coastal shorelines from California south to Peru, and from the se. Atlantic and Gulf coasts south to s. Atlantic coastal areas in South America.

Sandpipers and Allies (Family Scolopacidae)

Western Willets show notable physical differences between males and females. Two nonbreeding birds in Sebastian, Florida in late January show marked variation in body size, bill length, and leg length (up to 25% for all features) between the smaller male at left and larger female. A Ruddy Turnstone bathes in the foreground.

Not usually aggressive toward their own species while foraging in winter, a Western Willet at Fort Myers Beach, Florida in January decided to take matters into its own bill by spearing a competing bird in this squabble for feeding space. The competitor was trying to get the upper "hand" by landing on the back of the front bird. He chose poorly!

252

▲ When breeding season is over, Western Willets head for coastal areas where their preferred feeding habitat is the surf zone. This group of molting birds on the NJ shore in July is foraging in the wave zone for their favorite Atlantic Coast meal, mole crabs, which birds secure by probing into the shallow surf as waves retreat.

Interior migrants use marshes and lakeshores for feeding, while coastal migrants and wintering birds prefer rocky coastlines, sod banks, tidal flats, beaches, and shallow bays. Most juveniles fledge by mid-July and begin their southbound migration within a week or two. They arrive on the Pacific, Gulf, and Atlantic coasts in mid to late July and peak in late August.

Wintering birds in coastal zones can be quite aggressive toward other willets if they venture into their preferred feeding space and will spear their competitors with their long bills. Usually these confrontations end with one bird retreating, but the photo shown here paints a painful picture, with one bird's bill embedded in the chest of the other willet. Along the Pacific Coast in winter they join Marbled Godwit, Whimbrel, and Long-billed Curlew in roosting flocks and in the wave zone, foraging for crustaceans, mollusks, marine worms, and fish.

Kevin looked through an extended several-mile concentration of about 50,000 Western Willets in February 2012 on the mudflats outside Panama City, Panama, while searching for an Eastern Willet, but with no luck, attesting to the regional regularity of Eastern Willet's winter range. Population estimates for Western Willet remain at 160,000, with a total population of 250,000 birds including Eastern Willet (Morrison, et al. 2001). However, with Kevin estimating almost 50,000 Western Willets in a long stretch of beachfront mudflats in Panama in February 2012, it is easy to believe their population is substantially higher than the estimate from 2001. It's amazing what a little protection can do.

▶ Greater Yellowlegs
Tringa melanoleuca

BIOMETRICS 11½–13¼ inches long; wingspan: 28–29½ inches; weight: 11–235 grams (4–8.3 ounces)
STRUCTURE Rangy with long legs, neck, and bill; small head with Adam's apple in retracted neck
STATUS Common at interior and coastal sites during migration; fairly common along Atlantic, Pacific, and Gulf coastlines in winter. Population trends stable or slightly increasing

This large, rangy, sabre-billed wader is big enough to consort in deeper water with herons and egrets, though it generally shuns crowds and is often seen foraging alone or in loose small- to medium-sized groups. With a high-stepping gait, it strides through the shallows in search of invertebrate prey and small fish that it plucks from the water column with a stabbing thrust of its straight or upturned bill (females have longer bills that often turn upward near the tip).

Lesser Yellowlegs picks while Greater stabs, and birds are quite defensive about their favorite fishing holes. During migration, Greater Yellowlegs sometimes forms a line to drive small fish into the shallows, where easier picking occurs. More often than not, feeding Greater Yellowlegs are widely spaced when foraging.

These hyper-alert birds are quick to announce your intrusion with their loud, ringing, 3-note descending whistle that is easily imitated. Along with Black-bellied Plover and Killdeer, they are among the first shorebirds to take

It's a big, lanky *Tringa* sandpiper with flashy yellow legs, a long stride, and a loud voice. Often referred to as "marshpiper" because of its habit of wading in deeper water than other sandpipers, Greater Yellowlegs nests in buggy bogs in vast boreal forests from Alaska to Newfoundland. It is one of the most visible and widespread shorebirds in North, Central, and South America.

flight when intruders are sensed, and they have a nervous nature rather than a trusting one, like Lesser Yellowlegs. Since they regularly give a loud, ringing alarm call when taking flight, other shorebirds react to this and often follow in escape flight.

Compared with Lesser Yellowlegs, Greater has a rangier body shape that Kevin likens to a cowboy who has spent a great deal of time in the saddle. It also has an unbalanced, belly-heavy body shape with a very long neck that when retracted shows a distinct Adam's apple that is lacking in

This female Greater Yellowlegs with very long bill is ever alert and vocal when agitated, and you can almost hear the ringing *TU-tu-tu* call as it lifts up into flight. The old market gunners used to call Greater the "telltale," whose loud warning cries spoiled many a shot. TEXAS, MARCH

▲ Towering over the more delicately proportioned Lesser Yellowlegs, a Greater Yellowlegs (center) shows the longer legs; bulkier, rangy body shape; and longer neck with Adam's apple look created by its long, retracted neck. Lesser has a more gently rounded body profile with even lines and a thinner, more pointed bill. Birds are molting into breeding plumage on March 14 at High Island, Texas.

Lesser. Greater's legs are proportionally longer than Lesser, giving it a lanky appearance and a gawky walking gait. Greater walks with a jerky motion and often runs frantically to chase small fish (a behavior rarely seen in Lesser Yellowlegs). It also sweeps its bill side to side at the water surface, particularly in low light or in murky water.

The bird's vigilance and warning cries earned it the ill-spoken nickname "tell-tale" or "tattler" by market gunners, and likely as many birds were shot out of pique as were shot for profit. A famous painting by Philadelphia artist Thomas Eakins called *Whistling for Plover* depicts how this was done. The gunner, wearing a white shirt and cradling a single barrel shotgun, is seen crouched in open salt marsh with lips puckered and the crumpled forms of eight or nine birds around him. Upon close examination, the birds are not plovers, but yellowlegs, and while Greater Yellowlegs did not undergo the degree of annihilation suffered by many flocking shorebirds, including Lesser, Edward Howe Forbush nevertheless lamented in 1912 that Greater Yellowlegs had "diminished greatly."

Today the bird is a common and widespread migrant whose numbers are hard to assess owing to a very large

▽ Nonbreeding Greater Yellowlegs (*left*) shows brownish-gray upperparts with neat black and white notches, thin pale streaks on head and neck, and a yellowish to gray bill base. A crisp-plumaged 1st-winter bird (*right*) exhibits long, broad wings; a long, tubular body with white rump; a long, thick bill; and long legs whose feet extend well past the tail in flight. FLORIDA, JANUARY (BOTH IMAGES)

breeding distribution that spans boreal forest habitats from Newfoundland to Alaska. It also has a huge wintering range that extends south along both coasts from s. British Columbia and s. New England, as well as the s. United States and Caribbean south through the entirety of Central and South America, where birds use a range of freshwater and tidal habitats, including rice fields and mangroves.

Spring migration occurs between early February and early June, sooner than most other shorebirds. Leaving South American sites in February–March, birds gather by the thousands on the coast of Suriname. After a cross-Gulf migration, they make their way north in small loose flocks that may in times and places number 80–200 birds (notably the marshes of Delaware Bay, March–May). It is most common across the southern tier of states in April, and in northern states from late April to mid-May. Birds arrive on breeding grounds in late April in the West, and early to mid-May in the East.

Breeding farther south and earlier than most Nearctic shorebirds, Greater establishes widely spaced territories which are defined with frequent and vigorous flight displays that involve a repeated sequence of flapping, gliding, and stopping, resulting in an undulating flight that may last for 15 minutes while the bird sings. Birds place their 3–4 eggs in a shallow depression in moss, peat, or crowberry in the shade of a stunted tree or shrub in flat terrain close to water in muskeg forests. Leaving the nest within a day of hatching, young are tended to by both

Similar postures on these breeding yellowlegs in May in NJ show the longer legs and neck, larger size, and longer, thicker bill on Greater Yellowlegs (rear). Plumage on these early spring birds is very similar. A heavily marked breeding Greater Yellowlegs from Florida in April (right) gets ready to enjoy a fish dinner after snatching one from the water column.

Sharing muskeg breeding habitat with Mew Gull near Homer, Alaska in May, this Greater Yellowlegs shows complete, dramatic breeding plumage. In spring, legs may become orange-tinged due to elevated sexual hormones.

While plumage of yellowlegs can be quite similar, physical differences are not. Greater Yellowlegs' larger size, longer legs, longer, thicker bill, and rangier, bulkier body with Adam's apple look are obvious in this great comparison photo of juveniles in September in NYC. Lesser's more evenly rounded body shape is also evident. By standing on one leg, birds reduce the loss of body heat and also give the other leg a chance to rest.

adults until chicks are able to fly in 18–20 days. Evidence suggests that chicks range widely, even when young.

Southbound migration is protracted (late June–late November) and conducted on a broad front, with birds flying solo or in small- to medium-sized flocks. Adults begin to depart breeding grounds in late June (mostly females and failed breeders) and arrive in wetlands of North, Central, and n. South America by early July. Greater Yellowlegs tends not to join other shorebird flocks, but sometimes migrates alongside Lesser Yellowlegs. Birds are believed to use the same routes in spring and fall. Southbound juveniles depart the breeding grounds in mid to late July and arrive in n. South America in early October. Peak numbers of juveniles occur through much of North America in late September and October.

While population numbers are hard to assess because of the huge breeding range, migratory counts and Breeding Bird Surveys resulted in the previous estimate of 100,000 birds in 2001 (Morrison et al. 2001) to be raised to 137,000. Counts remain above 1970 levels, and the trend across all years (1974–2009) was an increase of 1.1% per year.

The extent of breeding distribution increased by 78% in Ontario between 1981–1985 and 2001–2005 (Cadman et al. 2007 in Andres et al. 2012), and counts in Suriname remained similar between 1970s–1980s and recent years (Ottema and Rancheran 2009 in Andres et al. 2012). Therefore, the population is most likely stable or slightly increasing (Andres et al. 2012).

Wilson's Phalarope
Phalaropus tricolor

BIOMETRICS 8¾–9½ inches long; wingspan: 15½–17¼ inches; weight: 30–85 grams (1–3 ounces)
STRUCTURE Plump, rounded body; small head and long, slim neck; medium-length, needle-thin bill; relatively short legs with webbed feet
STATUS Common, particularly inland; scarce in the East. Population is stable in the short term; possibly declining in the long term

Wilson's Phalarope is one of three members of this genus (Red and Red-necked phalaropes are the others) that spend a great deal of their lives swimming in shallow to deep waters where they feed on aquatic invertebrates. They are the only shorebirds to utilize vegetation-free aquatic habitats on a regular basis, and though they belong to the family Scolopacidae (the sandpipers), they are always listed as "allies" to this family group and rarely, if ever, referred to as sandpipers.

This needle-billed, gazelle-necked shorebird is less purely aquatic than the other two members of this genus that spend most of the year in offshore marine environments. Wilson's, however, does share with them the spinning feeding behavior that ferries food to the surface, where it is more easily plucked (see page 260, top photo). Microscopic organisms are drawn up along the bill by utilizing the surface tension that binds water droplets

▲ This amazing high-contrast photo of a breeding female Wilson's Phalarope by Scott Elowitz sets an ethereal mood with the fog and whitish water merging into one entity and is one of Kevin's favorites in the book. Note the very long, needle-like bill on this female bird, which differs greatly from the shorter bills of the other two phalarope species. NEBRASKA, MAY

together and traps organisms within. They also follow other waterbirds, including Blue-winged Teal and Northern Shoveler, to feed on nutrients that are stirred up. In shallow water, Wilson's may feed by rapidly scything their bill sideways through water, similar to Lesser Yellowlegs.

Wilson's also actively feeds on land, chasing and plucking prey from the mud or shallow water, often with head held low and neck outstretched. Their erratic movements on land often disturb other feeding shorebirds. Since they lack developed salt glands and the lobed toes of other phalaropes, they are rarely found at sea. They have webbed toes, but not lobes on the sides of the toes.

Northbound migration is largely overland and takes place from mid-March and early May. Northbound migrants depart wintering grounds in mid-March and move through the highlands of South America, from which they continue through Central America or across the Gulf of Mexico. Up to 90% of the breeding population passes through Cheyenne Bottoms NWR in Kansas in non-drought years, and birds reach breeding areas in late April and early May.

Wilson's Phalarope was once more numerous and widespread across wetlands of the western interior, but land use practices have resulted in substantial loss of wetlands,

▼ Odd body shapes and quirkish behavior while foraging on land set this "sandpiper" apart from all others. A bird at left with head and neck tilted down and tail raised is frantically and awkwardly chasing insects along the shoreline, a behavior that often "spooks" other shorebirds. At right, odd plump bodies and neckless profiles are often seen on relaxed birds. LOS FRESNOS, TEXAS, NOVEMBER

A male Wilson's Phalarope flies above his breeding territory where his chicks are hiding in tall grass near Logan, British Columbia in June. As a polyandrous shorebird, his plumage is duller than that of females, and he incubates eggs and broods young with no help from her. Females are also the dominant pair bond member.

and today Wilson's is a widespread breeder in upper western states and provinces, with smaller numbers in a narrow band across the Great Lakes region. They nest on the edge of a wetland, or on vegetated plains adjacent to marshes down to the size of roadside ditches. Breeding occurs early to late May, and nest sites, selected by the female, are typically located in concealing vegetation that is denser than that chosen by other prairie breeding shorebirds.

The male alone incubates 4 eggs for 23 days, and young leave the nest within 24 hours of hatching and can swim 1 hour after hatching. While highly aquatic, most of Wilson's activity on the breeding grounds involves walking and wading in shallow water, and their prey is mostly small aquatic invertebrates. Phalaropes have less fidelity to breeding areas than most other prairie shorebirds, perhaps to compensate for drought and mercurial water conditions.

Wilson's practices a polyandrous (many mates) mating system, which while not unique to the phalaropes does exhibit a degree of expression beyond the limits of other polyandrous shorebirds. In phalarope society, it is the more brightly colored female that initiates pair bonding and mating, leaving the male to incubate, hatch, and rear young by himself. As soon as young are able to fend for themselves, males join females at migratory staging areas. The female

Landing gear down! These Wilson's Phalaropes paint a pretty picture with their attractive breeding plumage and engaging reflection in Arizona in April. Peggy Wang captured this wonderful scene which shows two males in front and one third from the left, with the rest females.

typically deserts the male shortly after the last egg is laid and then often competes for other males. She later joins other females in hypersaline pre-migratory staging areas, where numbers may swell into the hundreds of thousands in mid-June through July at such key locations as Mono Lake in California and the Great Salt Lake, Utah.

At these locations, Wilson's feasts on an abundance of brine flies and brine shrimp. This gorging has to satisfy the energetic demands of feather replacement and the storage of fat reserves they will need to make the nonstop oceanic migration from staging areas to wintering grounds in the Andes of South America, where the birds will once again gather in hypersaline lakes and alkaline ponds.

Fall migration takes place from mid-June to early November, with adult females and failed breeders moving first and males about 2 weeks later. Juveniles move more slowly and across a broader range overland in w. and central US, with smaller numbers in e. United States, from which they continue through Mexico to n. South America. Southbound adults undertake a fast, nonstop flight of more than 50 hours from staging areas in w. North America to coastal w. South America, with almost no adult records from Central and n. South America. Adult birds reach South America in early August, and juveniles in late August. Adults and juveniles migrate south through the Andes, arriving at wintering areas from late September to early November.

Wilson's is one of the only shorebirds that molts at resting sites on the migratory pathway rather than on the breeding grounds before leaving, or on the wintering grounds (Cornell's All About Birds, "Wilson's Phalarope" 2023). While stopping to molt on salty lakes in the West, Wilson's may double their body weight after gorging on ample available food. Sometimes they put on so much weight that they cannot fly, allowing researchers to catch them by hand.

As a result of challenges posed by drought, the breeding range is expanding in all directions, and after population declines in the 1980s, numbers now appear stable in the short term. Estimated population numbers are about 1.5 million birds (Morrison et al. 2006).

◤ You can't see me" is what this nonbreeding Wilson's Phalarope is saying to the flies it is stalking by lying flat in shallow water with its neck and head lowered. This behavior is unnecessary as the flies don't even try to avoid the intense foraging behavior of Wilson's, but perhaps it makes the bird feel accomplished. TEXAS, NOVEMBER

◀ Almost fairy-like in gentle elegance, this doe-eyed juvenile migrant Wilson's Phalarope in late August in NYC has already replaced its dark-centered juvenile scapulars (upper back feathers) with fresh, pale gray nonbreeding ones as it swims while foraging in shallow water.

Two juvenile Wilson's Phalaropes show their deceptively large size and plump, football-shaped bodies when compared with an adult Semipalmated Sandpiper in August at Jamaica Bay Wildlife Refuge in Queens, NYC. Wilson's odd flight profile with long wings and very short tail/rear body is unique among shorebirds as compared with a Semipalmated Sandpiper below.

Large flocks of Wilson's Phalaropes stage at western locations like this one in Wilcox, AZ in early August prior to making long, migratory flights to the Andes high-elevation saline lakes in s. South America. Small numbers of birds also winter at southern locations in the United States. Wilson's Phalarope is found only in the Americas, unlike its two global phalarope relatives.

▶ Red-necked Phalarope
Phalaropus lobatus

BIOMETRICS 7¼–7½ inches long; wingspan: 12¾–16¼ inches; weight: 20–48 grams (0.7–1.7 ounces); larger than "peep" and smaller than Red Phalarope
STRUCTURE Compact body shape with slim neck, small head, needle-thin bill, and short legs with webbed, lobed toes
STATUS Common offshore, and locally common in the West; uncommon on Atlantic Coast in fall; population trends mostly unknown

Red-necked inhabits open water bodies, from small breeding ponds to large lakes and open oceans, and is the most abundant and widely distributed of all the phalaropes.

Mostly aquatic and constantly in motion, they feed by jabbing and picking tiny invertebrates from or just below the water's surface, gliding across the water in a restless, zigzag path somewhat reminiscent of storm petrels. They also spin in shallow water to create a vortex, which draws aquatic invertebrates to the surface (see page 262, top photo). Red-necked is more terrestrial than Red Phalarope, but less so than Wilson's, and they will sometimes walk at the water's edge while picking at flies and other insects. Red-necked flies rapidly, usually low to the water, and lights on the surface with amazing ease and quickness.

This smallest and most delicate of the phalaropes also ranks among the planet's smallest pelagic birds, spending up to 9 months per year at sea, coming ashore only to breed in the Arctic and subarctic regions of the planet.

Brightly colored breeding-plumaged female Red-necked Phalaropes rank among the most stunning of all shorebirds. All phalaropes practice polyandry, where the female is the dominant and most colorful member of the pair bond, and her only parental duty is to lay the eggs. After that, the male incubates the eggs and broods the chicks while she looks for another male to mate with. ALASKA, JUNE

In spring at the Great Salt Lake in Utah, tens of thousands of Red-necked Phalaropes stage and fatten up on flies and other invertebrates prior to completing their migration. Flying fast and low, this smallest of pelagic birds combines both action and grace in their mobile flocks. Can you spot the two obvious Wilson's Phalaropes in this flock? Hint: upperwings lack the bold white stripe. An immature California Gull stands in the shallow water. UTAH, MAY

They are typically seen in small, active flocks, but occasionally in large groups numbering in the tens of thousands at favored stopover or wintering sites. It is not unusual to see these hardy shorebirds 100 miles or more from the nearest coastline, where they feed on copepods and other marine invertebrates along lines of seaweed or other flotsam. They are truly at home in the open ocean, a harsh environment that prevents most other birds from living or even visiting there.

Spring migration occurs between mid-March and early June, primarily along the coast (western US) or offshore, though many fly overland. Migration also takes place through the interior West and offshore Atlantic, but the origin of birds off the Atlantic Coast is unknown. Peak migration takes place along the US Pacific Coast from mid-April to late May, and through the interior West and offshore Atlantic in May.

In North America, Red-necked breeds from w. Alaska east across n. Canada to southern Greenland. Arrival on the breeding grounds takes place from mid-May to early June, with females arriving first. Red-necked breeds in low Arctic and subarctic tundra as well as alpine tundra, and prime habitats are rich in freshwater ponds, pools, bogs, and marshes. Nests are situated on mounds, tussocks, or clumps of grass and sedges near the water's edge or surrounded by water. High breeding concentrations are associated with tundra habitats that contain ample polygonal formations.

Similar to other phalaropes, Red-necked practices polyandry at breeding locations, where the female is the dominant and most colorful of the sexes. As the initiator of the pair bond, females often get into fierce fights over males they want to mate with. After laying 3–4 eggs, females desert the males and look for another male to mate with, or they join other females in small groups to

△ With rugged, Precambrian outcrops in the background, more than 100,000 Red-necked Phalaropes stage at Great Salt Lake, Utah in May prior to migrating to Arctic and subarctic breeding grounds. Late summer draws even more numbers of this species and is truly a sight to behold.

▽ A female Red-necked Phalarope is completing her molt to breeding plumage in Homer, Alaska in May, where she might breed in this muskeg bog near the boreal forest. Note the webbed, lobed toes that allow Red-necked to navigate open oceans for up to nine months of the year. At right, a breeding male shows less gaudy plumage than females in Utah, May.

fatten up before migrating south. The male assumes all parental roles, including incubation, brooding, and protecting the young until they can fly.

While not territorial, Red-necked Phalaropes may sometimes form loose colonies, and birds are often polyandrous where there is an excess of males. Clutch size is typically 3–4 eggs, the first of which is laid in a bare scrape. Eggs are incubated for 18–21 days by the male alone, who adorns the nest with leaves and stems after the first egg is laid. Chicks can run, swim, and feed themselves within hours of hatching, with the male providing protection and warmth on cold Arctic days.

Young fledge in 16–17 days, and the male leads them to feeding areas rich in insects and midges. Females migrate first from mid-June to mid-July, shortly after egg laying, and attending males and failed breeders follow between early July and mid-August. Birds first move to migratory staging areas like Mono Lake, California, and the Great Salt Lake, Utah (where brine flies in assorted stages of development are the main food source). The Bay of Fundy is also a major staging area.

Migrant adults first appear along the Pacific Coast in late June/early July and peak in late July and August. Juveniles are more widespread and more regular than adults along the Atlantic Coast, with peak numbers occurring from early September to early October. Fall aggregations may number from many hundreds of thousands up to a million birds.

By early November, most birds have departed for wintering areas at sea in tropical waters where warm and cold currents meet and oceanographic features create an upwelling that brings tiny crustaceans (copepods) to the surface. They also concentrate where whales feed, and in the s. Atlantic are associated with Sargasso mats and the marine life that flourishes in these drifting aquacultures.

Most North American breeders gather in the waters of the Humboldt Current off w. South America, where birds spin to create their own micro-upwellings that bring plankton trapped in water droplets within reach of their bills. They then use the surface tension that binds the water droplets to guide them and their encapsulated prey to the back of their mouth by manipulating the droplets so they flow in the direction that minimizes surface tension.

Because of the wide breeding range of Red-necked Phalarope, coverage of Arctic PRISM surveys (Bart and Smith 2012) was too incomplete to revise the previous estimate of 2.5 million birds in North America (Morrison et al. 2006), with a global population of 4.1 million (Cornell's All About Birds, "Red-necked Phalarope" 2023). Besides the very large population declines in the Bay of Fundy in the 1990s, no reliable information on population trends is available.

◀ Mating while floating! Now that is truly unique among shorebirds. These Red-necked Phalaropes are copulating while floating on the water in Alaska in late May. Females will lay 3 or 4 eggs in about 5 days and then leave all the parenting to the male.

▼ Three tiny gnomes with feathers are Red-necked Phalarope chicks that just hatched, and within hours they will be out of the nest and feeding themselves. Attending males have to provide only warm feathers and protection against predators until natal down is replaced by juvenile feathers. With eggs safely tucked beneath their mates, female Red-neckeds may vie for another male or take a few weeks off by joining other females on tundra ponds before migrating to hypersaline lakes, like Great Salt Lake, Utah, where more than a million staging Red-necked Phalaropes may gather in July to feast on brine flies prior to migrating to open oceans. ALASKA, JUNE

Juvenile Red-necked Phalaropes don't take a back seat to adults when it comes to beauty. This young bird in August in NYC shows a pleasing mix of black, white, and gold feathers and an engaging elegance that captures the fancy of every birdwatcher. A trio of older juveniles at right in California in late August whose gold and brown feathers have faded still show a striking mix of black, white, and gray on the upperparts.

Red Phalarope
Phalaropus fulcarius

BIOMETRICS 8–8¾ inches long; wingspan: 16–17½ inches; weight: 36–77 grams (1.3–2.75 ounces)

STRUCTURE Bulkier than Red-necked with a heavier chest, thicker neck, larger head, longer wings and tail, and heavier bill; short legs with webbed, lobed toes

STATUS Common far offshore; rare nearshore or inland. Long-term population trends unknown, but changes at individual Arctic study sites indicate an apparent decline.

Highly pelagic in nature, Red Phalarope is one of the most strikingly plumaged breeding shorebirds, sporting a bright red body, white cheek, black cap and yellow bill. Spending most of its life at sea, this circumpolar Arctic breeder even migrates over marine environments. Clearly an aquatic species, Red Phalarope swims with amazing buoyancy, as if barely touching the water, and flies skillfully over rough surf, swirling high to avoid waves. It occasionally submerges its head or upends its body to secure prey.

With tightly packed breast feathers, the birds can sit on water like a duck, and given their nonbreeding gray upperparts, rafts of phalarope are nearly invisible until the sea sprouts wings. Known to whalers as the "whale bird," Red Phalaropes do indeed attend feeding whales, taking full advantage of the surfacing leviathan's self-generated upwelling to feast on zooplankton brought to the surface.

While common outside the breeding season, Red Phalaropes are seldom found in nearshore waters or on land. Migrants are regularly seen in deep water off both coasts and are also regular along the Pacific Coast in fall. Wintering in marine environments in small to very large flocks of several thousand birds, Red Phalaropes are concentrated mostly along thermal fronts or upwellings in the waters off South America and Africa where water density changes and zooplankton are concentrated. Red Phalarope is generally found farther offshore than Red-necked

Breeding female Red Phalarope is one of the most stunning, colorful shorebirds in the world. A feeling of calmness pervades her expression while she sits in a tundra pond in Prudhoe Bay, Alaska in June. Note the dark, wet brood patch on her breast, which means she just finished her time sitting on the recently laid eggs. Now it is up to the male to take over the rest of the parental duties.

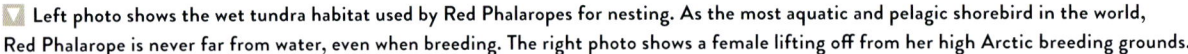
Spending most of their lives at sea, these nonbreeding Red Phalaropes were photographed off California's coast in December by Mike Danzenbaker. Note how they ride buoyantly on the surface of the water as though pumped up with air, unlike Red-necked Phalaropes, which ride low in the water like submarines.

Left photo shows the wet tundra habitat used by Red Phalaropes for nesting. As the most aquatic and pelagic shorebird in the world, Red Phalarope is never far from water, even when breeding. The right photo shows a female lifting off from her high Arctic breeding grounds.

Phalarope and may consume smaller invertebrates, but they sometimes form mixed flocks with Red-necked, especially far offshore.

The convergence zones frequented by Red Phalarope are usually visible because of surface slicks, feeding whales, and concentrations of fish, which all attract the birds. In typical phalarope fashion, birds spin in the water, creating a vortex rich in plankton, which birds pluck from the water column, sometimes upending in the process. Papillae in the bill are used to strain prey from the water. On the breeding grounds, Red Phalaropes feed mostly on crustaceans and insects, especially midges and flies.

Spring migration occurs between mid-March and early June, primarily along the coast in the West or offshore, though many birds fly overland. Arrival on breeding grounds takes place from mid-May to early June. Breeding generally farther north than Red-necked Phalarope and closer to coasts, Red Phalarope nests near ponds in marshy polygonal tundra dominated by sedges. In phalarope fashion, Red Phalaropes practice reverse sexual dimorphism, or polyandry, with females larger and more brightly plumaged than males.

Pair bonding is initiated by either sex and features spirited aerial chases that may involve multiple birds. Though mostly monogamous after pair formation, females may have multiple mates when available, and some failed nesting males may pair with different females. Birds do not establish territories, but pairs defend the small area surrounding themselves and their mate, with females taking the lead role (see bottom photo on this page). Feeding is often done communally, or solitary birds may join groups that can number in the hundreds on ponds or at sea. On breeding grounds, birds also forage on foot, wading in shallow water like other shorebirds, or walking on land while picking prey from the surface of the mud.

Timing of nest construction is related to snow cover, with earliest snow-free areas in Alaska in early June, and in the far North by mid to late June. Nest scrapes precede eggs by a day or two. Both sexes participate in nest site selection, though it is unknown which sex makes the final determination. Females typically depart as soon as the last of 4 eggs are laid, which the male alone incubates for 19 days. In years of late snow cover, multiple pairs may be clustered in a small area.

Red Phalarope is a long-distance migrant between Arctic breeding grounds and temperate to tropical oceans, mostly over open pelagic waters. Fall migration takes place between late June and early January, with adult females, failed breeders, and nonbreeders starting movements to sea in late June. Remaining males depart mid-July to early August, and juveniles depart breeding areas in early August and stage in nearshore areas until early September before continuing out to sea. First arrival of juveniles off the Pacific Coast occurs in late August and

▲ Mating Red Phalaropes show the duller-plumaged male engaging with a beautiful female on the Alaskan tundra in June. Right photo reveals a male incubating eggs on a nest that is well concealed by grasses, rendering the bird almost invisible without careful scrutiny.

▼ As polyandrous species, female phalaropes choose their mates and are serious protectors against intruders. A mated pair of Red Phalaropes at left shows the female watching another male trying to steal her away from her mate. She responds by attacking in midair as she chases him from her loose territory and established pair bond while her male swims quietly and passively away. ALASKA, JUNE

peaks from early to mid-September to late October, with occasional large movements into early December.

Migration is almost wholly offshore (up to 100 miles) in flocks of 100 birds or more, with some individuals already south of the equator by August. Most are present on winter territories by November. Principal wintering areas are well off the coast of w. and sw. Africa (for birds breeding in e. Canada) and off s. South America in the Pacific, where birds concentrate in the waters of the Humboldt Current.

Limited Arctic PRISM surveys (Bart and Smith 2012) produced an estimate of 1,617,000 birds in North America, and this number will likely increase with the completion of the Arctic PRISM surveys (Andres et al. 2012). No information on long-term trends exists, but individual Arctic study sites indicate a decline.

La Niña weather systems can inflict great losses on wintering Red Phalaropes in some years, which results in substantial short-term declines in breeding numbers, which Kevin experienced in his fourth year as a shorebird biologist in Alaska in 1995. Red Phalarope numbers declined almost 40% that year compared with numbers from the same location in 1994. This decline was due to a La Niña weather pattern in the s. Pacific, where warming ocean surface temperatures prevented the upwelling of cold currents that carry food to the surface. This resulted in the large percentage of mortality in Red Phalarope numbers that year, and a substantial decrease in breeding numbers. The species, however, shows an uncanny ability to bounce back from these population declines in a relatively short time.

▲ While sporting a duller breeding plumage than females, a male Red Phalarope from Alaska in June (*left*) still paints an attractive picture in his breeding attire. A fairly fresh juvenile (*right*) from Cape May, NJ in mid-August has already replaced its juvenile back feathers with nonbreeding ones. Many juveniles partially molt prior to leaving breeding areas where they hatched, usually involving nonessential body and upperpart feathers.

▼ A moody shot of a female Red Phalarope flying over the Arctic tundra was taken by Ted Swem while working as a biologist for the US Fish and Wildlife Service in Alaska. The beauty of this image goes beyond words! ALASKA, JUNE

Jacanas (Family Jacanidae)

As adroit as a pickpocket in a crowd, jacanas walk across lily pads, plucking an insect or snail here or peeling back the lip of a lily pad there. Jacanas are the unrivaled masters when it comes to exploiting the riches of this food-rich stratum. Flycatchers can nibble at the edges, herons wade into the gunnels, but in deeper water away from land, jacanas reign supreme.

These rail-like shorebirds of tropical and subtropical regions occur in extensive freshwater wetlands with emergent vegetation (lettuce wetlands), and all eight global species are flamboyantly plumaged and noisy. Not found in Europe or across much of North America, Northern Jacana is the only species that occurs in the Greater Antilles, Mexico, and most of Central America, and it ranges far enough north to reach the lower Rio Grande Valley of Texas, where it is a rare but regular vagrant.

The feet of jacanas have exquisitely long toes and nails that distribute the birds' weight across lily pads and mats of other aquatic plants, allowing jacanas almost exclusive access to the food riches of these environs and conferring the ability to walk on water, a specialization that has earned it the name "Jesus bird" in Jamaica. The name

Jacana derives from the Native American name *naha'na*, which Carl Linnaeus, the father of present-day taxonomy, translated somewhat imperfectly into Latin.

While weak fliers, jacanas are strong swimmers, despite their ungainly feet, and they dive as quickly as grebes, remaining underwater for extended periods with just the tip of their bills visible. Young also dive to escape danger. Birds can be very vocal at times, emitting an array of sounds that range from piping and twittering to cackles and rattles. The cackle call is most often given as birds take flight.

Jacana society is polyandrous bordering on promiscuous, with females mating simultaneously with up to four males who fully assume the role of parenting, from nest building to incubation to brooding. As befits shorebirds, young jacanas feed themselves, but adults do accompany

Can this really be a shorebird?! The answer is still yes, but wait for more advanced DNA research that might change this designation. Using exquisitely long toes to distribute its weight across lily pads, this Northern Jacana appears to be walking on water. Northern Jacana does not mix with other shorebirds, but it does associate with Purple Gallinule and Common Moorhen in their shared tall-grass marsh habitat. COZUMEL, MEXICO, FEBRUARY

chicks and lead them to food, and females vigorously defend their territory against other females.

Eggs are deposited on floating mats of vegetation, and nest platforms are flimsy and typically wet, as the bulk of the platform is below water. A typical clutch is 4 eggs, and the eggs tend to be smaller relative to the female's size than those of other shorebirds. Clutch loss runs high in jacanas, justifying the polyandrous mating strategy, and unlike other shorebirds, chicks develop slowly, with first flight occurring in 6–12 weeks. Juveniles may remain with the male for up to 12 weeks, and Northern Jacana adolescents can stay in the parents' territory until molting into adult plumage at about 12 months of age. With cold weather not a concern in these tropical areas, there is no need to hurry.

While no jacana species is threatened, their dependence on clean, healthy wetlands places them at risk of filled or drained marshes for agricultural purposes, including rice plantations. The introduction of non-native aquatic plants that crowd out water lilies is also problematic for this family. However, jacanas readily adapt to shallow artificial wetlands and botanical parks that showcase ponds with lily pads.

▶ Northern Jacana
Jacana spinosa

BIOMETRICS   6¾–9½ inches long; wingspan: 17–22 inches; weight: 82–161 grams (2.9–5.75 ounces)
STRUCTURE   Chunky body, small head and thin neck; long legs with extremely long toes; slightly larger than Greater Yellowlegs
STATUS   Locally common Mexican and Central American species; irregular visitor to s. Texas (mostly October to May); casual to s. Arizona

- -

Festively colored with a chestnut body and black head and neck that sets off the bird's bright yellow bill and forehead, these dedicated marsh birds habitually raise their wings, flashing bright yellow flight feathers. In flight, the long greenish legs trail behind.

While Northern Jacana is genetically placed in the shorebird family Jacanidae, a sighting of these birds creeping along the edges of a shallow water body or walking nimbly on lily pads may remind you of a Purple Gallinule because of their behavior and their incredibly long toes at the end of

◁ Adult Northern Jacana is festively colored with chestnut wings and back and black head, neck, and underparts that are complimented by a bright yellow bill and forehead. Birds walk with a slow, deliberate gait as they navigate floating vegetation in quiet ponds in tropical areas. BELIZE, FEBRUARY

Flashing yellow flight feathers that contrast with a rich chestnut body and underwing coverts, a male Northern Jacana protects his young chick. As a polyandrous species, females mate with multiple males and lay eggs, but they do not participate in any parental duties other than incubating their eggs until a full clutch is laid. Note the bright yellow spiked spur on the forewing, used in combat in days gone by and still used by African Jacanas. BELIZE, FEBRUARY

long, spindly legs. They walk on floating vegetation while picking at plants or the surface of the water and may use their long toes to flip over lily pads. Their preferred habitats include freshwater marshes with abundant floating vegetation, flooded grassy fields, and roadside ditches.

Food includes insects, seeds, aquatic vegetation, small fish, and aquatic invertebrates, which they pick off root balls. They walk with a slow, deliberate gait, often lifting their wings to maintain balance. Northern Jacana does not typically fraternize with other shorebirds but does associate with Purple Gallinule and Common Moorhen.

Breeding occurs seasonally or year-round, depending on the permanence of the marsh. Incubation is 22–24

days, and the polyandrous female may mate with up to four males per season. Each male may raise two or more broods, and he alone incubates the eggs and cares for young. Young can fly after about 57 days.

These mostly resident birds range from Mexico south along both slopes of Central America to Panama and the Greater Antilles. While they are year-round residents where appropriate habitat exists, some birds move locally to more permanent ponds when drought or dry seasons reduce water levels. They are rare and irregular visitors to s. Texas, where they have bred, and a casual visitor to s. Arizona. Juveniles account for about 75% of US vagrant records.

SECTION 3
Epilogue

Normally a placid species, this male Sanderling is aggravated by another Sanderling who wandered too close to its feeding territory during migration in St. Augustine, Florida in May. Its back feathers are raised as a warning to the encroaching bird to stay away, and its body language is one of aggression.

Shorebird Populations: Going Down?

We were complacent too long. After the International Migratory Bird Act banned market gunning in 1918, the Western Hemisphere's depleted shorebird numbers began to rebound, their populations no longer suppressed by unsustainable losses to their breeding populations. We presumed that given protection, numbers would simply continue to climb, and even those species whose numbers had failed to recover quickly to pre-market gunning levels (Pectoral Sandpiper, Buff-breasted Sandpiper, and Upland Sandpiper) would restore themselves in time, like the American Golden-Plover.

And numbers did indeed climb throughout much of the 20th century, but then the pendulum swung, populations began to falter, and there were early signs in the latter half of the 20th century that some shorebird populations were in distress. But it was the rapid decline of northbound Red Knots on the beaches of Delaware Bay in the 1990s (a decline caused by overharvesting of horseshoe crabs, whose eggs the knots depend on) that demonstrated the fragile nature of shorebird recovery and survival.

Shorebirds live on a knife's edge, with their habitat needs specific and their migratory timetables and windows for breeding fairly exacting. If they miss the seasonal abundance of food resources at key migratory stopover sites, the birds will lack the fuel to reach the Arctic in condition to breed, or they may not complete the journey. Even more damning are changes wrought on the Arctic biome by climate change. Indeed, most of the shorebird species whose populations are declining are Arctic breeders. The overall decline in North America's shorebirds since 1973 is approaching 40% (Learn 2022), a staggering figure. And while climate change is widely considered to have reduced nesting success, climate-related changes to the Arctic biome are not the only challenge faced by these pan-hemispheric denizens, and for some species, challenges away from the breeding grounds may be the most determining.

North America has seen a loss or degradation of 50% of its wetlands in the last 300 years (Pyle 2019) with California having lost 90% of wetlands overall (Yaich 2017). Climate change also affects critical coastal wetlands where inland marsh expansion has failed to keep pace with rising sea levels. Exacerbating the challenge to shorebirds is coastal habitat loss, caused by development and bulkheading.

Other challenges include changing agricultural practices and application of pesticides that accumulate in wetlands and flooded rice fields in the Mississippi Delta, reducing aquatic insects and occasionally poisoning

▼ Horseshoe crab eggs are a vital food resource for migrating Red Knots, Ruddy Turnstones, Sanderlings, and Semipalmated Sandpipers on the Delaware Bay in May. The serious over-harvest from 1990–2005 led to a severe decline in Red Knot numbers, but thanks to conservation efforts, a comeback is in progress.

△ During migration, shorebirds need the latitude to feed interrupted for long periods. Disruptions to their feeding pattern, however innocent or benign, prevent them from putting on weight and contribute to a decline in their fat reserves. It could also compromise the success of their migration.

feeding shorebirds (KK, pers. comm.). Recreational use of beaches where shorebirds winter in both the Northern and Southern Hemispheres can force beach-feeding shorebirds to flush repeatedly, and these cumulative disruptions may be as harmful to birds trying to meet energy needs as outright habitat loss.

Were you to travel to Morro Bay State Park in California in winter, you would find numbers of Marbled Godwit, Long-billed Curlew, Whimbrel, and Western Willet whose longer bills and legs allow them to exploit the food riches of the standing water lying between the high tide line and the Pacific Ocean. Compounding the problem are the large numbers of beach strollers in this zone and board surfers in the surf zone. Of smaller shorebirds like Sanderling and Dunlin, there are none, their feeding zone usurped by beach strollers and dog walkers.

While shorebirds are protected in North America (except for American Woodcock and Wilson's Snipe, which are still hunted), they are still legally hunted for sport in the Caribbean and netted for food in South America. It is estimated that 12,000–34,000 shorebirds are killed annually for "sport" hunting on Barbados alone. There are no estimates for shorebirds netted for food in n. South America and China, but the numbers are very large and increasing every year.

While shorebirds are mostly immune to predation by roaming cats, a leading cause of songbird decline, both songbirds and shorebirds are seined out of the air by utility lines that flank feeding areas. And beach-nesting shorebirds like Piping and Snowy Plover are ever challenged by mammalian and aerial predators like gulls, crows, foxes, raccoons, and skunks attracted to food left on beaches by human visitors. Whatever the cause or causes of the current shorebird decline, it is almost certain to have at its core one or more human-related drivers. That's the bad news. The good news is that if humans are the cause, we are also the solution.

What You Can Do

With shorebird populations imperiled, the question naturally becomes: "What can the average citizen do to support a shorebird survival turnaround"?

Perhaps most important is for citizens of the United States and Canada to support their respective fish and wildlife services. Staffed by dedicated biologists standing on the front line of habitat and migratory bird protection, the U.S. Fish and Wildlife Service maintains 560 national wildlife refuges encompassing 31 million acres which are largely wetlands that benefit migrating and breeding ducks and shorebirds, as well as other wildlife. In Canada, 51 key locations covering 2.1 million acres of federally owned land are designated National Wildlife Areas.

It is incumbent on citizens of both the United States and Canada to support politicians and policies that advocate for fully funded and staffed fish and wildlife services.

$15 VOID AFTER JUNE 30, 1993

MIGRATORY BIRD HUNTING AND CONSERVATION STAMP

SPECTACLED EIDER

U.S. DEPARTMENT OF THE INTERIOR

◁ The purchase of Federal Duck stamps is a direct link to habitat protection, and a way that every conservation-minded individual can contribute to the preservation of waterfowl and other bird families.

Also important is opposition to any legislation or policies contrary to the dedicated purpose of wildlife refuges, which protect the plants and animals that constitute every citizen's natural dowry.

A tangible way of demonstrating support of shorebird habitat is the purchase of a Federal Duck Stamp, a mechanism signed into law by Franklin Delano Roosevelt during the Dust Bowl era for the purchase and enhancement of wetlands. At a modest $25 and available from the U.S. Postal Service and National Wildlife Refuges, duck stamps directly contribute to more and better wetlands for breeding and migrating shorebirds.

While Federal Duck Stamps do not support wetlands procurement outside US borders, Ducks Unlimited is a nonprofit organization active in wetlands protection on all sides of the border (north and south), which helps to bridge the gap. Ducks Unlimited and several other organizations work on behalf of shorebird protection and thus deserve your support. These include Manomet Bird Observatory, the National Audubon Society and regional chapters, and The Nature Conservancy.

On the local level, municipal land use practices can advantage migrating shorebirds. In many municipalities, sewage treatment facilities integrate shorebird-friendly water holding impoundments into the treatment process, a boon to shorebirds migrating over arid and forested regions, and they also provide ideal viewing areas for birdwatchers.

On a personal level, when vacationing on coastal beaches, be mindful of the needs of breeding and feeding shorebirds. Keep dogs on leashes and perhaps walk higher up on the beach to avoid flushing concentrations of feeding birds at the water's edge. Obey warning signs delineating shorebird nesting areas. These are also important roosting areas for wintering shorebirds.

Most of all, take action to reduce your carbon footprint, starting with making one less trip to the supermarket per week or investing in an electric car or truck. Such restraint can greatly reduce CO_2 emissions, helping to mitigate the global impact of climate change, whose disruptive influences are disproportionately felt in the Arctic, where most of our shorebirds breed, creating earlier snow melt and a mismatch between chick development and insect availability.

Critical Shorebird Habitats

Shorebirds are not equally distributed, and there are some key locations in North and South America that at times support 15–30% of the flyway's population of the Western Hemisphere's migrating and wintering shorebirds. These are part of the Western Hemisphere Shorebird Reserve Network (WHSRN), a powerful and influential global organization that deserves your contributions. Locations include the following:

- Copper River Delta, Alaska
- Bay of Fundy, Nova Scotia
- San Francisco Bay, California
- Stillwater NWR, Nevada
- Great Salt Lake, Utah
- Mono Lake, California
- Cheyenne Bottoms, Kansas
- Quivera NWR, Kansas
- Bolivar Flats, Texas
- Delaware Bay, Delaware/New Jersey
- Maryland-Virginia Barrier Islands
- Estero Rio Colorado, Mexico
- Narismas Nacionales, Mexico
- Marinhão, Brazil
- Paracas National Reserve, Peru
- Laguna Mar Chiquita, Argentina
- Lagoa do Peixe National Park, Brazil
- Chiloé Island, Chile
- Tierra del Fuego, Chile and Argentina

Needless to say, the protection of these key shorebird concentration areas is paramount in our international efforts to preserve shorebird species whose annual cycles span hemispheres.

Shorebirds are truly world citizens, and therefore every nation's responsibility.

AND YOURS.

North American Shorebird Population Estimates, 2018

A reassessment of population sizes and trends for all shorebirds that occur in North America has produced higher estimates for 28 of 71 populations and lower estimates for 7 populations compared with estimates by Andres et al. 2012. The numbers matter because they are put to many uses, including setting target population sizes, evaluating the effectiveness of conservation efforts, and identifying Important Bird Areas and wetlands of importance.

Biologist Brad A. Andres, national coordinator of the U.S. Shorebird Conservation Plan, prepared the latest estimates along with scientists from Environment Canada and the Manomet Center for Conservation Science. Most of the increases, he writes, were the result of more comprehensive surveys or re-analyses of existing data, not actual increases. Only four populations truly grew after the estimates of Morrison et al. (2006): the Great Lakes and Great Plains populations of Piping Plover, Hawaiian Black-necked Stilt, and Upland Sandpiper.

Analysis of data from migration counts and other sources revealed that the proportion of species exhibiting

▶ The winsome Piping Plover leads a precarious existence in the mercurial world between the waves and sand. While its numbers have increased over the last 30 years, rising sea level will surely be the demise of the Atlantic Coast subspecies.

increasing, decreasing, or stable long-term population trends has not changed much since 2006. But for many shorebird migrants in eastern North America, declines in the 1980s and early 1990s appear to have been followed by stable or increasing numbers.

"Still, the conservation status of North American shorebirds warrants concern," Andres warns. Consistent declines across all survey methods and time periods are evident in Snowy Plover, Killdeer, Mountain Plover, Lesser Yellowlegs, Whimbrel, Ruddy Turnstone, Red Knot, Sanderling, Semipalmated Sandpiper, Western Sandpiper, Pectoral Sandpiper, Purple Sandpiper, and Dunlin. Even worse, a pervasive lack of monitoring data means "we still have virtually no indication of the population trend for 25% of the shorebird taxa breeding in North America."

Who Were the Market Gunners?

They weren't bad men or evil men. They were hard-working watermen who took their living from the bounty of the natural world and directed their efforts toward whatever the season had to offer.

In winter, it was the rafts of waterfowl that carpeted the Chesapeake or the back bays of New Jersey. In March and April, these seasoned outdoorsmen dropped nets in Delaware Bay for Shad, or netted Alewives running in New England streams that connected freshwater lakes to the sea.

After a winter of salt cod, fresh oily smelt and a plate of fried shad roe was a treat to the palette. In May, the great hosts of northbound shorebirds began flooding the unglaciated marshes that stretched from Hoboken, New Jersey to Charleston, South Carolina, and in June, nets were stretched again for Striped Bass and Bluefish. Then back to the marshes from July to November for the southbound exodus of shorebirds. Next came rail birds and dwindling numbers of waterfowl. The ancestors of these Baymen once hunted Atlantic Right Whales in dories launched from shore, but the whales in Delaware Bay were hunted out and gone, and their numbers have never recovered.

Come December, when the gales might pin these hardiest of men indoors for several days, they gathered around wood stoves glowing with heat. Here they carved shorebird decoys and "stool" or replacement heads for decoys so disfigured by shot that they would not fool even a Dunlin, known to the gunners as "the Simpleton." As they carved, they reminisced about the vast clouds of shorebirds they had known in their youths. Where now there were scores, then there had been thousands. These were men like Captain Jessie Birdsall from Barnegat, Harry M. Shourds from Atlantic City, Captain Willy Sutton from Goshen, and Walker Hand from Cape May. Baymen and watermen, striving to make a living so their families might enjoy a better standard of living.

There are none alive today who witnessed the great hosts of birds. What attest to those halcyon days are the artfully crafted decoys, fashioned from knives whetted down to slivers, that are called "folk art" today and command prices that would make these old watermen gasp, more money at auction for a single carved bird than any would make in a lifetime. But to the watermen, the decoys were tools of their trade, which in concert with their rusted 10- and 12-gauge doubles put bread on the table, if not money in the bank.

Market gunning was challenging but not particularly lucrative. Red Knot sold for a dime apiece in market. Of course, the gunners who sold their birds by the barrel got only a fraction of that because there were shipping costs, middlemen and market sellers who also needed to make a profit, not to mention the cost of shot and powder. But killing shorebirds for market is what they did because it was their birthright and their trade.

Year by year, they saw the dwindling numbers and knew deep in their souls that the Passenger Pigeon's demise was a portent of what was to come, but it was precisely the loss of the pigeon that made shorebird gunning profitable. The nation's industrial centers were filling with European immigrants hungry for an inexpensive source of protein. Migrating shorebirds filled the bill. It was food on the table not only for watermen, but also for mill workers in Five Points and the wealthy patrons of New York's finest eating establishments. Everyone ate shorebirds until, barrel-full by barrel-full, the inexhaustible numbers were depleted.

Still the watermen plied their trade, because it was all they knew, and because their families needed food on the table. They'd rise early to catch an incoming tide, hitch up the horses, and drive their rigs close to the marshes, leaving the horses in the shade and perhaps wrapped in tar-soaked burlap to give them some protection from the evil, biting greenhead flies. Once there, fathers and sons would haul their decoys and guns out to the edge of a shallow salt marsh pond where shorebirds appeared to be dropping in yesterday.

▶ A soft, serene mood surrounds this resting juvenile Semipalmated Sandpiper in Queens, NYC in August. Ever vigilant, even at the young age of two months, this crisp-plumaged youngster keeps one eye open for potential danger.

"Maybe, today," the seasoned gunners mused. "A day like the good old days when clouds of plovers would drop among the decoys."

"Certainly, today," the younger gunners dreamed as they set out the decoys, inhaling the tang of salt-laden air and watching the sun winking over the barrier islands. It felt good to be a man, doing a man's work and learning the tricks of the trade from a master waterman, dad.

"Set them decoys in a U or a Y," the old man admonished, "so the birds can feel comfortable landing between the prongs and we won't have to shoot among the stool."

"Yes, sir," the boy affirmed.

Then with the decoys spread, the gunners crouched near the water's edge and waited for whatever the rising tide and their falling fortunes might bring. Waiting is what hunting is mostly about. As they waited, watching the distant clouds of birds rising over the flats as the incoming tide pushed them closer, the sun began to heat the air, and the wings of the biting greenhead flies took on a shrill quality. Straw hats offered some protection from the sun and the flies, but the wash-thinned, sweat-compressed flannel shirts were no defense against the scissors-like mandibles of the greenhead flies. Before midmorning, the shirts of man and boy were blotched with blood. It was hard not to flinch as the fangs of greenheads sank into tender flesh.

"Steady now," the older gunner cautioned. "I see them," the boy acknowledged. A flock of dowitchers led by a Black-bellied Plover was coming their way. "They'll come in from the right to land into the wind," the old man promised. "Right," the boy breathed.

And with the tight-packed flock setting their wings, he started to rise. *I can't miss*, he thought.

"Steady now. Wait for them to land," the old man counseled.

And land they did. Right in the fork of the Y.

Without rising, the old man fired both barrels. The boy, too.

Shot dimpled the water and birds fell, some dead, some wounded. The survivors took wing, whistling their lament, *tu-tu-tu...tu tu tu*, taps for their fallen comrades.

"Should I collect the birds?"

"No," the old man said. The wounded will just attract more birds. "Keep a low profile when you reload," he cautioned.

Yes, shorebirds were sometimes shot before they landed, especially if it was evident that they were spooked.

But this wasn't sport hunting, this was market gunning on an industrial scale, and the more birds you could kill with a single load of shot, the more profitable it was. The next flock to come in was Dunlin, or "Simpletons." "Don't shoot," the old man whispered. "They're not worth the price of powder. There's more Robin Snipe [dowitchers] heading this way."

By day's end, they had enough birds to almost fill two peach baskets. Mostly dowitchers, but also several Black-bellied Plovers, or "bullheads," and a Greater Yellowlegs, or "telltale," that landed among the fallen birds and, sensing danger, started calling loudly, drawing the old man's ire and finally the muzzle of his gun. "Waste of good shot and powder," the old man said aloud. But that "telltale" was going to ruin their day. When the tide began to recede, the birds went back to feeding on the flats and stopped flying.

The old man collected the decoys while the boy gathered the birds. "Enough for two peach baskets. Pretty good," the boy said.

▽ Heads up! Short-billed Dowitchers were a favorite target for the market gunners along the Atlantic Coast in the 1800s and early 1900s as they tended to fly in flocks and were large enough to provide ample food.

But the old man, keeping his thoughts to himself, mused that *times were, we'd fill six baskets*.

"You did real good," was all he said, then added, "Now the work starts. Maybe after tomorrow we'll have enough to drive up to Atlantic City, you up for that?"

"Yes, sir. Does that mean I can skip my chores?"

"It does not. See if you can find one or three birds fit for dinner, but not the plovers, they'll fetch more at the market."

▶ The market gunners called Greater Yellowlegs "Tell-tale," because its loud warning cries and nervous nature caused incoming shorebirds to divert their path and flare out of the hunter's range. It was despised by the gunners because of this and usually only shot out of spite.

▼ This foraging Black-bellied Plover appears to have the Midas touch. Mostly solitary feeders, Black-bellieds largely escaped the decimation market gunning levied upon American Golden-Plover (a flocking species).

Acknowledgments

Pete

In setting out to write a brief acknowledgment, I realize such will not be the case. The fact is that many have added the weight of their experience to the hand on the author's pen, most notably mentors Floyd P. Wolfarth and Richard Kane, plus birding companions Don Freiday, Peter Bacinski, Clay Sutton, and David Sibley. We are grateful to shorebird biologists Charles Duncan and David Mizrahi for insights reflected in these pages as well as to Brian Harrington and Larry Niles, whose May banding projects on the shores of Delaware Bay conferred hands-on experience.

It would be an indiscretion beyond bearing not to acknowledge the inspiration of this book's predecessor *The Shorebirds of North America* (1967) by Peter Matthiessen, Gardner D. Stout, Robert Verity Clem, and Ralph S. Palmer. And every book written by me in the last 40 years has been moved forward by the skill of my incomparable agent, Russell Galen. The appeal of this book is due in no small part to Princeton University Press's Robert Kirk, an editor whose support is underscored by the book's visual appeal.

"As many images as you like," was Kirk's counsel to photographer Kevin Karlson, who, even as coauthor, deserves special acclaim. Readers cannot begin to appreciate the many thousands of hours spent gathering the array of photos housed in this book, nor the hundreds of hours spent in the lightroom grooming those images to a polished luster. We are especially indebted to my good friend and retired fish and wildlife biologist Ted Swem of Fairbanks, Alaska, whose contributions to the photo array elevated the book's standing and appeal.

As always, I am indebted to wife Linda, who figures in every endeavor, and to my good friends Dorothy Claire for editorial assistance and Beth VanVleck, whose writer's retreat on the shores of Penobscot Bay provided the perfect blend of comfort and solitude. Beth, your gifted photo of E. B. White was ever looking over my left shoulder, and his editorial eye passed judgment on every word.

Kevin

So many people have influenced my birding and wildlife photography life and career, but to keep it short, I would like to acknowledge a few people who have played a major part in my passion for shorebirds.

Paul Buckley was one of the earliest influences in my study of shorebirds, with his knowledge and attention to detail providing a good foundation. Tom Davis from Queens, NY, set a bar of excellence with shorebird ID in the early 1980s that inspired me to study and know this wonderful bird family. Don Riepe, from the Jamaica Bay Wildlife Refuge in Queens, provided friendship, a house to crash at in Howard Beach, NY, while I photographed shorebirds at the Refuge, and maintains a loyal bond that continues to this day, not to mention easy access and guidance to this superb shorebird location. Declan Troy from Anchorage, Alaska, hired me to work on the North Slope of Alaska as a shorebird biologist from 1992 to 1995 and taught me the value of proper scientific protocol when doing research on shorebirds in high Arctic locations.

As a shorebird photographer, I started out shooting side by side with Arthur Morris at Jamaica Bay, originally from Brooklyn and Queens, NYC, and now Florida. Arthur set a high bar of excellence with bird photography, and for many years we shared our photos and techniques, which helped make me the professional wildlife photographer that I am today. A number of his wonderful images also appear in this book. Other photographers that had positive influences on my work include Jim Zipp, Greg Downing, Lloyd Spitalnik, Scott Elowitz, Alan Murphy, Milo Burcham, Kevin Laughlin, Rob Shepherd,

◁ A small juvenile Semipalmated Sandpiper feels protected and comfortable enough to catch a few winks in front of a large juvenile Greater Yellowlegs at the Jamaica Bay Wildlife Refuge in NYC in September. Shorebirds perch on one leg when tired to take weight off the other leg for a little while.

Tim Grey, Stacey Sather, Mike Danzenbaker, Greg Lasley, and Johan Schumacher. So many other shooters have influenced my work, but space does not allow all my thanks to them here.

Another shoutout to Declan Troy of Troy Ecological Research Associates, who gave me the opportunity as a shorebird biologist during four summers on Alaska's North Slope to study and photograph more than 20 species of breeding North American shorebirds, with many of those photos appearing in this book. And to echo Pete Dunne, we are indebted to Ted Swem of Fairbanks, Alaska, who provided us with unique and unmatched shorebird photos from Arctic habitats in the Y-K Delta and North Slope of Alaska taken during his 35 years as a biologist with the U.S. Fish and Wildlife Service.

I would also like to pay tribute to the many shorebird biologists I had the pleasure of learning from and working with on the North Slope of Alaska, including Craig Hohenburger, Russell Fraker, Tim Menard, Sophie Webb, Nils Warnock, Dave Rudholm, Katie Duffy, Laura Payne, and Jeff Davis. I would also be remiss if I did not honor my two coauthors of *The Shorebird Guide* (2006), Michael O'Brien and Richard Crossley, who shared their incredible knowledge about shorebirds with me during this project.

More tributes to the many shorebird biologists and researchers conducting important research on Red Knot, Ruddy Turnstone, Sanderling, and Semipalmated Sandpiper on the Delaware Bay from 1992 to present, led by Larry Niles and Amanda Dey. Some of these include Clive Minton, Humphrey Sitters, Mark Peck, Guy Morrison, Charles Duncan, Brian Harrington, Dick Veitch, Patricia González, Joe Jehl, Kathy Clark, Joanna Burger, Jane Goletto, Eric Stiles, Sherry Meyer, Patti Hodgetts, and many more. This groundbreaking work included horseshoe crab harvest and relating it to declining shorebirds which rely on their eggs, and much of it is referenced throughout our book. While Dr. David Mizrahi, vice-president of research at the NJ Audubon Society, also worked alongside the Bayshore researchers, he has been a friend and colleague since the 1990s, and his groundbreaking work on Semipalmated Sandpiper, including banding and weighing more than 45,000 birds since the early 1990s, has contributed to the needed reference sources on this declining species.

Special thanks to my friend and colleague from Alaska, Rick Lanctot, who allowed me to photograph Buff-breasted Sandpiper breeding behavior on his study plot in 1992 on Alaska's North Slope, and whose postdoctoral work and knowledge of this species is unsurpassed in the world. As Buff-Breasted Sandpiper is my favorite bird species of all, this experience stands out among many shorebird encounters on the Slope.

Finally, to go back to the beginning of my birding career, I would like to thank Bob Perna for taking my fiancée, Dale Rosselet, and me to Sandy Hook, NJ in 1978 to "look at birds." We had no interest in doing this, but joined him anyway, and the seed was planted. A trip to the Everglades the following winter cemented my connection to the amazing world of birds, and 45 years later, it is still the sun my world revolves around.

Last but not least, thanks and appreciation to my editor and close friend at Princeton University Press, Robert Kirk, whose guidance, trust, and superb knowledge of how to produce a high-quality, interesting book make him the most important person in these acknowledgments. Thank you from the bottom of my heart, Robert. Special thanks also to David Price-Goodfellow at D & N Publishing for overseeing the book's excellent design and to Natalie Baan, senior production editor at Princeton University Press, for her superb final editing of the entire book.

TESTIMONY FOR PHOTOGRAPHY IMAGES

Many images in this book depict shorebird behavior and nest defense that might cause some people to think that the photographers were disturbing birds on the nesting grounds. However, Kevin and Ted Swem, who took all of these behavioral Arctic images, worked as biologists in Alaska where their work required finding and checking nests of breeding shorebirds. Therefore, these images were taken while performing required biological tasks for important research and not for personal gain.

Eastern Willet is a large, boisterous shorebird that breeds along Atlantic and Gulf coastlines and winters in coastal South America. Kevin took this photo of a bird at Bolivar Flats, Texas in April with its landing gear down.

Appendix

Rare Shorebird Vagrants

The focus of this section is 16 species of shorebirds whose appearance as vagrants in North America is intermittent (less than annual), as well as those species whose occurrence, albeit regular, is limited to the continent's geographic extremities like the islands off the coast of w. Alaska. Species for which there are fewer than 10 North American records are determined to be accidental and are not included here. Our list of casual species, where the word "casual" means "very rare," is shown here.

NORTHERN LAPWING *Vanellus vanellus*—This rakishly crested Eurasian breeder is a casual visitor in late fall and winter (AOU Checklist) along the Eastern Seaboard and inland to Ohio. During winters when deep snowpack covers inland Europe, late winter movements of birds seeking open foraging areas in Britain and Ireland may result in modest numbers of birds overshooting and being propelled by strong easterly winds to e. North America. Incursions involving multiple birds occurred in the winters of 1927 and 1966 (Howell et al. 2014). The single spring record in Newfoundland also coincided with easterly winds.

EUROPEAN GOLDEN-PLOVER *Pluvialis apricaria*— A round-bodied tundra breeder from n. Europe, Russia and Iceland, the bird is "almost annual" in spring in Newfoundland, usually in small flocks (Howell et al. 2014). Very few records exist away from Newfoundland, even during spring seasons subjected to influxes of this species, which may number in the thousands (MacTavish in Howell et al.).

COMMON GREENSHANK *Tringa nebularia*—This large, lanky Tringa breeds in taiga forest clearings and boggy forest habitats from northern Scotland to Kamchatka and winters from Spain and Africa to se. Asia and Australia. It is, in many respects, the Old World counterpart to Greater Yellowlegs. In spring, it is a rare but fairly regular migrant in the w. Aleutians as well as in Alaska's Pribilof Islands. While a vagrant to Iceland, there are numerous fall and winter records from the Azores to Barbados (Howell et al. 2014).

SPOTTED REDSHANK *Tringa erythropus*—This lanky Tringa breeds on shrubby tundra and taiga edge from Scandinavia to e. Siberia and is a rare but regular mostly fall migrant on the w. Aleutians and Bering Sea islands. This species also shows a broader pattern of vagrancy, mostly along the Atlantic coast from Newfoundland south to Barbados, as well as multiple records from British Columbia to s. California.

WOOD SANDPIPER *Tringa glareola*—Closely related to Solitary Sandpiper, this boreal forest breeder is found across n. Eurasia from the UK to Kamchatka and is an uncommon but regular spring visitor to the western and central Aleutians where it has bred (Kessel 1989).

GREEN SANDPIPER *Tringa ocrophus*—Breeding in the boreal forests of Eurasia, this medium-sized sandpiper is a casual spring visitor to Iceland and the outer Aleutians.

GRAY-TAILED TATTLER *Tringa brevipes*—A rare and irregular spring migrant to the outer and central Aleutian and Bering Sea Islands, it is uncommon in fall and casual elsewhere in coastal Alaska. Also known as Polynesian Tattler, this species is second only to Sharp-tailed Sandpiper in terms of fall encounters on Bering Sea Islands, with both species breeding in similar taiga zone habitats and wintering on islands in the South Pacific (Howell et al. 2014).

COMMON SANDPIPER *Actitis hypoleucos*—Sometimes considered conspecific with Spotted Sandpiper, this sandpiper of Eurasia is a rare but regular spring migrant through the w. Aleutians and rare in fall, and the bird has bred on Attu (Howell et al. 2014).

TEREK SANDPIPER *Xenus cinereus*—This frenetic mudflat and shoreline feeder is a rare and irregular mostly spring migrant through the w. Aleutians and a casual summer visitor to the islands of the Bering Sea (Kessel 1989).

FAR EASTERN CURLEW *Numenius madascariensis*—This forceps-billed, e. Asia breeder is casual in spring and early summer on the Aleutians and Pribilofs and in w. Alaska. Most birds winter coastally in Australia, making it an ideal candidate for vagrancy toward w. Alaska.

BLACK-TAILED GODWIT *Limosa limosa*—Black-tailed breeds in wet grasslands from Iceland to e. Europe and in pockets from e. Mongolia to Siberia and ne. China. In winter in estuaries and lagoons, it occurs in Ireland, w. France, and Portugal as well as Africa east to India, se. Asia and Australia. In spring migration, it is a rare but regular visitor to the w. Aleutians and casual on Bering Sea Islands. It is rare in spring in Newfoundland and very rare elsewhere along the Atlantic Seaboard.

GREAT KNOT *Calidris tenuirostris*—Breeding on montane tundra mostly in ne. Siberia, and wintering coastally in se. Asia, in spring this species is a very rare to casual migrant on the Aleutians and w. Alaska, and very rare on St. Lawrence Island and the Seward Peninsula.

LITTLE STINT *Calidris minuta*—As a rare visitor to North America with an arctic breeding range extending from Scandinavia to north-central Siberia, Little Stint winters mostly in Africa and India. Nevertheless, it is casual to accidental in the western to central Aleutians and Bering Sea Islands, mostly in fall, and shows a wide pattern of vagrancy along both North American seacoasts.

TEMMINICK'S STINT *Calidris temminicki*—Breeding in Arctic regions from Scandinavia to eastern Siberia and wintering in Africa and southeast Asia, it is an uncommon and intermittent migrant in the w. Aleutians and Bering Sea islands, and the rarest stint to occur in the rest of North America. A high count of 43 birds on Attu on May 23, 1991 (Howell et al. 2014) is considered exceptional. Typically the birds are found as individuals or in small groups.

LONG-TOED STINT *Calidris subminuta*—Breeding in open subarctic boreal habitats, mostly in Siberia, and wintering in Australasia, Long-toed is an uncommon and irregular spring migrant on the w. Aleutians and rare in fall (Howell et al. 2014). It is a very rare spring migrant on the islands of the Bering Sea (Kessel 1989).

SHARP-TAILED SANDPIPER *Calidris acuminata*—Breeding on grassy tundra across n. Russia and Siberia, Sharp-tailed winters from West Africa to Australia in a variety of coastal and inland wetlands. In spring, it is a regular migrant through the w. Aleutians and a rare spring migrant on the Seward Peninsula, where it has also been recorded in summer. In fall, it is a mostly uncommon but regular migrant in w. Alaska, where it may be locally common. Sharp-tailed is also a rare but regular fall vagrant along the West Coast south to California, casual on the Atlantic Coast, and very rare elsewhere, but it does show a widespread pattern of vagrancy.

In addition to these 16 species, there are 10 shorebirds whose occurrence in North America, while verified, is considered by the authors to be accidental (recorded fewer than 10 times). These include two tropical New World species, the Double-striped Thick-knee, *Burhinus bistriatus*, and Collared Plover, *Charadrius collaris*. Rounding out the list of accidentals are such Eurasian strays as Marsh Sandpiper, Common Redshank, Slender-billed Curlew, Eurasian Curlew, Spoon-billed Sandpiper, Broad-billed Sandpiper, European Woodcock, and Oriental Pratincole.

Bibliography

The following publications provided many of the facts and insights that enliven the pages of this book. In addition, assorted online resources were consulted, most notably the species accounts provided by the Cornell Lab of Ornithology at birdsoftheworld.org.

Alaska Shorebird Group. 2008. Alaska Shorebird Conservation Plan. Version II. Anchorage: Alaska Shorebird Group. Available at http://www.arlis.org.

American Ornithologist's Union. 1983. The AOU Checklist of North American Birds, 6th edition. Lawrence, KS: Allen Press.

Andres, B. A., and G. A. Falxa. 1995. The Black Oystercatcher (*Haematopus bachmani*). In The Birds of North America, No. 155 (A. Poole and F. Gill, eds.). Philadelphia: Academy of Natural Sciences.

Andres, B. A., P. A. Smith, R. I. G. Morrison, C. L. Gratto-Trevor, S. C. Brown, and C. A. Friis. 2012. Population estimates of North American shorebirds, 2012. Wader Study Group Bulletin 119:178–94.

Andres, B. A., J. A. Johnson, J. Valenzuela, R.I.G. Morrison, L. A. Espinosa, and R. K. Ross. 2009. Estimating eastern Pacific coast populations of Whimbrels and Hudsonian Godwits, with an emphasis on Chiloé Island, Chile. Waterbirds 32(2): 216–224.

Audubon, John J. 1827. Birds of North America. London: Havel Engravers.

Bent, Arthur Cleveland. 1962. Life Histories of North American Shorebirds, Parts 1 and 2. New York: Dover. First published 1927.

birdfact.com. "Curlew." https://birdfact.com/birds/curlew

birdfact.com. "Whimbrel." https://birdfact.com/birds/whimbrel

BirdLife International. 2023. Species factsheet: *Calidris ferruginea*. http://datazone.birdlife.org/species/factsheet/curlew-sandpiper-calidris-ferruginea

Boissoneault, Lorraine. 2021. "Piping Plovers: Despite new challenges, the birds make their comeback." Great Lakes Now. PBS, June 23, 2021. https://www.greatlakesnow.org/2021/06/piping-plovers-recovery-population/

Boyle, William J., Jr. 2011. The Birds of New Jersey: Status and Distribution. Princeton, NJ: Princeton University Press.

Carle, Ryan D., Gabbie Burns, Mary Clapp, Kayla Caruso, Deborah House, Ron Larson, Ashli Lewis, Ann E. McKellar, John Neill, Michael Prather, John Reuland, and Margaret Rubega. 2022. Coordinated phalarope surveys at western North American staging sites, 2019–2021. Unpublished report of the International Phalarope Working Group. https://doi.org/10.13140/RG.2.2.18546.17608

Colwell, M. A., and J. R. Jehl Jr. 1994. Wilson's Phalarope (*Phalaropus tricolor*). In The Birds of North America, No. 83 (A. Poole and F. Gill, eds.). Philadelphia: Academy of Natural Sciences.

Cooper, J. M. 1994. Least Sandpiper (*Calidris minutilla*). In The Birds of North America, No. 115 (A. Poole and F. Gill, eds.). Philadelphia: Academy of Natural Sciences.

Cooper, T.R., and K. Parker. 2011. American woodcock population status, 2011. U.S. Fish and Wildlife Service, Laurel, MD. 17 pp.

Corbat, C. A., and P. W. Bergstrom. 2000. Wilson's Plover (*Charadrius wilsonia*). In The Birds of North America, No. 516 (A. Poole and F. Gill, eds.). Philadelphia: Birds of North America.

Cornell Lab of Ornithology. 2023. "Red-necked Phalarope." All About Birds. Cornell Lab of Ornithology, Ithaca, NY. https://www.allaboutbirds.org/guide/Red-necked_Phalarope

Cornell Lab of Ornithology. 2023. "Ruddy Turnstone." All About Birds. Cornell Lab of Ornithology, Ithaca, NY. https://www.allaboutbirds.org/guide/Ruddy_Turnstone

Cornell Lab of Ornithology. 2023. "Wilson's Phalarope." All About Birds. Cornell Lab of Ornithology, Ithaca, NY. https://www.allaboutbirds.org/guide/Wilsons_Phalarope/

Del Hoyo, Josep, et al. 1996. Handbook of the Birds of the World, Vol. 3. Barcelona: Lynx Editions.

◀ This Semipalmated Sandpiper displays angel-like wings as it hover-feeds on mats of algae and mud. Long scimitar-shaped wings enable shorebirds to migrate long distances. NEW YORK CITY, AUGUST

Dinsmore, Stephen. 1994. Upland Sandpiper. In The Birds of North America (A. Poole and F. Gill, eds.). Philadelphia: Academy of Natural Sciences.

Dugger, B. D., and K. M. Dugger. 2002. Long-billed Curlew (*Numenius americanus*). In The Birds of North America, No. 628 (A. Poole and F. Gill, eds.). Philadelphia: Birds of North America.

Dunne, Pete, 2006. The Essential Field Guide Companion. New York: Mariner Books.

Elphick, C. S., and J. Klima. 2002. Hudsonian Godwit (*Limosa haemastica*). In The Birds of North America, No. 629 (A. Poole and F. Gill, eds.). Philadelphia: Birds of North America.

Elphick, C. S., and T. L. Tibbitts. 1998. Greater Yellowlegs (*Tringa melanoleuca*). In The Birds of North America, No. 355 (A. Poole and F. Gill, eds.). Philadelphia: Birds of North America.

Forbush, E. H. 1912. Nature Leaflet on Birds. Massachusetts Dept. of Agriculture.

Gatto-Trevor, C. L. 2000. Marbled Godwit (*Limosa fedoa*). In The Birds of North America, No. 492 (A. Poole and F. Gill, eds.). Philadelphia: Birds of North America.

Gatto-Trevor, C. L. 1992. Semipalmated Sandpiper. In The Birds of North America, No. 6 (A. Poole and F. Gill, eds.). Philadelphia: Academy of Natural Sciences.

Gill, Frank B. 1995. Ornithology, 2nd ed. New York: W.H. Freeman.

Gill, F., D. Donsker, and P. Rasmussen (eds.). 2021. IOC World Bird List (v11.2). https://doi.org/10.14344/IOC.ML.11.2

Gill, R. E., P. S Tomkovich, and B. J. McCaffery. 2002. Rock Sandpiper (*Calidris ptilocnemis*). In The Birds of North America, No. 686 (A. Poole and F. Gill, eds.). Philadelphia: Birds of North America.

Gill, R. E., Jr., P. Canevari, and E. H. Iversen. 1998. Eskimo Curlew (*Numenius borealis*). In The Birds of North America, No. 347 (A. Poole and F. Gill, eds.). Philadelphia: Birds of North America.

Griscom, Ludlow. Papers, 1918–1961. Cornell University Library, Collection Number 2701. Ithaca, NY.

Haig, S. M. 1992. Piping Plover (*Pluvialis melodus*). In The Birds of North America, No. 2 (A. Poole, P. Settenheim, and F. Gill, eds.). Philadelphia: Academy of Natural Sciences.

Hall, Henry Marion. 1960. A Gathering of Shore Birds. New York: Devin-Adair.

Handel, C. M., and R. E. Gill. 2001. Black Turnstone (*Arenaria melanocephala*). In The Birds of North America, No. 585 (A. Poole and F. Gill, eds.). Philadelphia: Birds of North America.

Harrington, B. A. 2001. Red Knot (*Calidris canutus*). In The Birds of North America, No. 563 (A. Poole and F. Gill, eds.). Philadelphia: Birds of North America.

Heiser, E., and C. Davis. 2020. Piping Plover nesting results in New Jersey: 2020. NJ Division of Fish and Wildlife Endangered and Nongame Species Program. https://www.nj.gov/dep/fgw/ensp/pdf/plover20.pdf

Holmes, R. T., and A. Pitelka. 1998. Pectoral Sandpiper (*Calidris melanotos*). In The Birds of North America, No. 348 (A. Poole and F. Gill, eds.). Philadelphia: Birds of North America.

Houston, C. S., and D. E. Bowen, Jr. 2001. Upland Sandpiper (*Bartramia longicauda*). In The Birds of North America, No. 580 (A. Poole and F. Gill, eds.). Philadelphia: Birds of North America.

Houston, C. S., C. Jackson, and D. E. Bowen, Jr. 2011. Upland Sandpiper (*Bartramia longicauda*), version 2.0. In The Birds of North America (P. G. Rodewald, ed.). Cornell Ithaca, NY.

Howell, Steve N. G., Ian Lewington, and Will Russell. 2014. Rare Birds of North America. Princeton, NJ: Princeton University Press.

Jackson, B.J.S., and J. A. Jackson. 2000. Killdeer (*Charadrius vociferous*). In The Birds of North America, No. 517 (A. Poole and F. Gill, eds.). Philadelphia: Birds of North America.

Jehl, J. R., Jr., J. Klima, and R. E. Harris. 2001. Short-billed Dowitcher (*Limnodromus griseus*). In The Birds of North America, No. 564 (A. Poole and F. Gill, eds.). Philadelphia: Birds of North America.

Jenni, D. A., and T. R. Mace. 1999. Northern Jacana (*Jacana spinosa*). In The Birds of North America, No. 467 (A. Poole and F. Gill, eds.). Philadelphia: Birds of North America.

Johnson, O. W., and P. Connors. 1996. American Golden-Plover (*Pluvialis dominica*), Pacific Golden-Plover (*Pluvialis fulva*). In The Birds of North America, No. 2001–2002 (A. Poole and F. Gill, eds.). Philadelphia: Academy of Natural Sciences.

Keppie, D. M., and R. M. Whiting, Jr. 1994. American Woodcock (*Scolopax minor*). In The Birds of North America, No. 100 (A. Poole and F. Gill, eds.). Philadelphia: Academy of Natural Sciences.

Kessel, Brina. 1989. Birds of the Seward Peninsula, Alaska. Fairbanks: University of Alaska Press.

Klima, J., and J. R. Jehl, Jr. 1998. Stilt Sandpiper (*Calidris himantopus*). In The Birds of North America, No. 341 (A. Poole and F. Gill, eds.). Philadelphia: Birds of North America.

Knopf, F. L. 1996. Mountain Plover (*Charadrius montanus*). In The Birds of North America, No. 211 (A. Poole and F. Gill, eds.). Philadelphia: Academy of Natural Sciences.

Lanctot, R. B., and C. D. Laredo. 1994. Buff-breasted Sandpiper (*Tryngites subruficollis*). In The Birds of North America, No. 91 (A. Poole and F. Gill, eds.). Philadelphia: Academy of Natural Sciences.

Learn, Joshua Rapp. 2022. "Wild Cam: Harvest may contribute to shorebird decline." The Wildlife Society, Jan. 28, 2022. https://wildlife.org/wild-cam-harvest-may-contribute-to-shorebird-decline/

Lowther, P. E., H. D. Douglas III, and C. L. Gratto-Trevor. 2001. Willet (*Catoptrophorus semipalmatus*). In The Birds of North America, No. 579 (A. Poole and F. Gill, eds.). Philadelphia: Birds of North America.

MacWirter, B., P. Austin-Smith Jr., and D. Kroodsma. 2002. Sanderling (*Calidris alba*). In The Birds of North America, No. 653 (A. Poole and F. Gill, eds.). Philadelphia: Birds of North America.

Marchant, J., T. Prater, and P. Hayman. 1986. Shorebirds: An Identification Guide. Boston: Houghton Mifflin.

Marks, J. S., T. L. Tibbitts, and R. E. McCaffery. 2002. Bristle-thighed Curlew (*Numenius tahitiensis*). In The Birds of North America, No. 705 (A. Poole and F. Gill, eds.). Philadelphia: Birds of North America.

McCaffery, B., and R. Gill. 2001. Bar-tailed Godwit (*Limosa lapponica*). In The Birds of North America, No. 581 (A. Poole and F. Gill, eds.). Philadelphia: Birds of North America.

McPeake, R., and J. Aycock. 2000. Migratory Stopover Habitat for Shorebirds. University of Alabama, Division of Agriculture, leaflet FSA 9108. https://www.uaex.uada.edu/publications/PDF/FSA-9108.pdf

Morrison, R.I.G., B. J. McCaffery, R. E. Gill, S. K. Skagen, S. L. Jones, G. W. Page, C. L. Gratto-Trevor, and B. A. Andres. 2006. Population estimates of North American shorebirds, 2006. Wader Study Group Bulletin 111: 67–85.

Morrison, R.I.G., R. E. Gill Jr., B. A. Harrington, S. Skagen, G. W. Page, C. L. Gratto-Trevor, and S. M. Haig. 2001. Estimates of shorebird populations in North America. Canadian Wildlife Service Occasional Paper No. 104. Ottawa.

Morrison, R.I.G., R. E. Gill Jr., B. A. Harrington, S. Skagen, G. W. Page, C. L. Gratto-Trevor, and S. M. Haig. 2000. Population estimates of Nearctic shorebirds. Waterbirds 23: 337–352.

Moskoff, W. 1995. Solitary Sandpiper (*Tringa solitaria*). In The Birds of North America, No. 156 (A. Poole and F. Gill, eds.). Philadelphia: Academy of Natural Sciences.

Moskoff, W., and R. Montgomerie. 2002. Baird's Sandpiper (*Calidris bairdii*). In The Birds of North America, No. 661 (A. Poole and F. Gill, eds.). Philadelphia: Birds of North America.

Mullarney, Killian, et. al. 1999. Birds of Europe. Princeton, NJ: Princeton University Press.

Muller, H. 1999. Common Snipe (*Gallinago gallinago*). In The Birds of North America, No. 417 (A. Poole and F. Gill, eds.). Philadelphia: Birds of North America.

Nettleship, D. N. 2000. Ruddy Turnstone (*Arenaria interpes*). In The Birds of North America, No. 537 (A. Poole and F. Gill, eds.). Philadelphia: Birds of North America.

Nol, E., and M. S. Blanken. 1999. Semipalmated Sandpiper (*Charadrius semipalmatus*). In The Birds of North America, No. 444 (A. Poole and F. Gill, eds.). Philadelphia: Birds of North America.

Nol, E., and R. C. Humphrey. 1994. American Oystercatcher (*Haematopus palliatus*). In The Birds of North America, No. 82 (A. Poole and F. Gill, eds.). Philadelphia: Academy of Natural Sciences.

O'Brien, Michael, Richard Crossley, and Kevin Karlson. 2006. The Shorebird Guide. New York: Houghton Mifflin.

Oring, L. W., E. M. Gray, and J. M. Reed. 1997. The Spotted Sandpiper (*Actitis macularia*). In The Birds of North America, No. 289 (A. Poole and F. Gill, eds.). Philadelphia: Academy of Natural Sciences.

Page, G. W., J. S. Warriner, and P. W. Paton. 1995. Snowy Plover (*Charadrius alexandrinus*). In The Birds of North America, No. 154 (A. Poole and F. Gill, eds.). Philadelphia: Academy of Natural Sciences.

Parmelee, D. F. 1992. White-rumped Sandpiper (*Calidris fuscicollis*). In The Birds of North America, No. 29 (A. Poole and F. Gill, eds.). Philadelphia: Academy of Natural Sciences.

Paulson, Dennis. 2005. Shorebirds of North America: The Photographic Guide. Princeton, NJ: Princeton University Press.

Paulson, D. R. 1995. Black-bellied Plover (*Pluvialis squatarola*). In The Birds of North America, No. 186 (A. Poole and F. Gill, eds.). Philadelphia: Academy of Natural Sciences.

Paulson, Dennis. 1993. Shorebirds of the Pacific Northwest. Seattle: University of Washington Press.

Payne, L. X., and E. P. Pierce. 2002. Purple Sandpiper (*Calidris maritima*). In The Birds of North America, No. 706 (A. Poole and F. Gill, eds.). Philadelphia: Birds of North America.

Peterson, Roger Tory. 1939. A Field Guide to the Birds. Second Edition. Boston: Houghton Mifflin.

Plauny, Holly L. 2000. Shorebirds. Fish and Wildlife Habitat Leaflet No. 17. Wildlife Habitat Management Institute and Wildlife Habitat Council, USDA. https://permanent.fdlp.gov/lps18594/www.ms.nrcs.usda.gov/whmi/pdf/shorebird.pdf

Pyle, Cooper. 2019. "Wetland Destruction in the United States." Planet Forward, Dec. 5, 2019. https://planetforward.org/story/wetland-destruction-united-states/

Readfearn, Graham. 2022, Oct. 26. Bar-tailed godwit sets world record with 13,560km continuous flight from Alaska to southern Australia. The Guardian.

Robinson, J. A., J. M Reed, J. P. Skorupa, and L. W. Oring. 1999. Black-necked Stilt (*Himantopus mexicanus*). In Birds of North America, No. 449 (A. Poole and F. Gill, eds.). Philadelphia: Birds of North America.

Robinson, J. A., L. W. Oring, J. P. Skorupa, and R. Boettcher. 1997. American Avocet (*Recurvirostra americana*). In The Birds of North America, No. 275 (A. Poole and F. Gill, eds.). Philadelphia: Academy of Natural Sciences.

Rubega, M. A., and D. M. Tracy. 2000. Red-necked Phalarope (*Phalaropus lobatus*). In The Birds of North America, No. 538 (A. Poole and F. Gill, eds.). Philadelphia: Birds of North America.

Sanders, F. J., M. C. Martin, M. D. Spinks, and N. J. Wallover. 2012. Abundance and distribution on Wilson's Plovers during the breeding season in South Carolina. The Chat 76: 117–124.

Senner, S. E., and B. J McCaffery. 1997. Surfbird (*Aphriza virgata*). In The Birds of North America, No. 266 (A. Poole and F. Gill, eds.). Philadelphia: Academy of Natural Sciences.

Sibley, David Allen. 2000. The Sibley Guide to Birds. New York: Knopf.

Skeel, M. A., and E. P. Mallory. 1996. Whimbrel (*Numenius phaeopus*). In The Birds of North America, No. 219 (A. Poole and F. Gill, eds.). Philadelphia: Academy of Natural Sciences.

Stone, Witmer. 1937. Bird Studies at Old Cape May. Philadelphia: Delaware Valley Ornithological Club.

Stout, Gardner D., Peter Matthiessen, Robert Verity Clem, and Ralph S. Palmer. 1967. The Shorebirds of North America. New York: Viking.

Svensson, Lars, and Peter J. Grant. 1999. Collins Bird Guide. London: Harper Collins.

Takekawa, J. Y., and N. Warnock. 2000. Long-billed Dowitcher (*Limodromus scolopaceus*). In The Birds of North America, No. 493 (A. Poole and F. Gill, eds.). Philadelphia: Birds of North America.

Tibbitts, T. L., and W. Moskoff. 1999. Lesser Yellowlegs (*Tringa flavipes*). In The Birds of North America, No. 427 (A. Poole and F. Gill, eds.). Philadelphia: Birds of North America.

Tracy, D. M., D. Schamel, and J. Dale. 2002. Red Phalarope (*Phalaropus fulicarius*). In The Birds of North America, No. 698 (A. Poole and F. Gill, eds.). Philadelphia: Birds of North America.

Warnock, N. D., and R. E. Gill. 1996. Dunlin (*Calidris alpina*). In The Birds of North America, No. 203 (A. Poole and F. Gill, eds.). Philadelphia: Academy of Natural Sciences.

Watts, B., and F. Smith. 2014. Hudsonian Godwits Go Long. Center for Conservation Biology.

WHSRN (Western Hemisphere Shorebird Reserve Network). 2021

Wilson, Alexander. 1818–1829. American Ornithology, 2nd ed. Philadelphia.

Wilson, W. H. 1994. Western Sandpiper (*Calidris mauri*). In The Birds of North America, No. 90 (A. Poole and F. Gill, eds.). Philadelphia: Academy of Natural Sciences.

Yaich, Scott, 2017. Defenders of Wildlife Magazine

Zöckler, Christoph. 2002. Declining Ruff *Philomachus pugnax* populations: A response to global warming? Wader Study Group Bulletin 97: 19–29.

▶ Mountain Plovers have declined dramatically over the last 70 years with conversion of Western prairies to agricultural land, so this juvenile "prairie ghost" from Sebastian, Texas in November represents hope for future generations and is a fitting way to end this book.

Photographer Credits

All photos by Kevin T. Karlson except for the 76 photos listed here:

Jamie Cunningham: page 89, upper photo; page 90, lower photo; page 96; page 128, three upper photos; page 174, lower photo; page 184, lower photo; page 191, lower photo

Mike Danzenbaker: page 70, right photo; page 71; page 92, full page; page 101, upper photo; page 124, both lower photos; page 134, upper photo; page 135, top right photo; page 158, lower left photo; page 159, lower photo; page 237, lower left photo; page 238, lower photo; page 265, upper right photo; page 266, upper photo

Linda Dunne: page 115, lower right photo

Scott Elowitz: page 196, upper photo; page 258

Brian Guzzetti: page 44; page 179, lower photo

Julian Hough: page 137, lower right photo; page 195; page 220, lower photo

Arthur Morris: page 62, lower photo; page 133; page 142, top photo; page 177; page 178, lower photo

Anita North: frontispiece, full page photo

David Speiser: page 127, upper photo

Lloyd Spitalnik: page vi, full page; page 41, lower photo; page 87, lower photo

Brian Sullivan: page 84, lower photo; page 91, upper photo; page 179, both upper photos

Ted Swem: page 3, lower photo; page 23, lower photo; page 24, full page; page 26, lower photo; page 39, top photo; page 81, right photo; page 86, upper photo; page 134, lower photo; page 135, top left photo; page 148; page 149, top photo; page 150; page 151, top photo; page 160; page 161, lower photo; page 200, upper photo; page 210, upper photo; page 219, upper right photo; page 220, upper left photo; page 233, upper and lower photos; page 234, lower photo; page 235, upper photo; page 264, upper photo; page 266, lower right photo; page 267, upper 2 photos; page 268, lower photo

Peggy Wang: page 259, lower photo

Sophie Webb: page 27, upper illustration

Audrey Whitlock: page 90, middle right photo